Automated Forms Processing

A Primer

How to Capture Paper Forms Electronically and Extract the Data Automatically

Published by CMP Books
an imprint of CMP, Media In c.
12 West 21 St
New York, NY 10010

ISBN 1-57820-049-0

Manufactured in the United States of America

First Edition, June 1996
Cover Illustration by by Saúl Roldán. Cover design by Mara Leonardi

Automated Forms Processing

A Primer

HOW TO CAPTURE PAPER FORMS ELECTRONICALLY
AND EXTRACT THE DATA AUTOMATICALLY

BY
HARVEY SPENCER

Table of Contents

Case Studies

AUTOMATED FORMS PROCESSING

Introduction to Automated Forms Processing

Why do we need Forms Processing?

Paper and electronic forms are the predominant way to transact business. Setting up, printing, completing and dealing with the forms is a major clerical task currently costing corporations billions of dollars per year. After design and printing, they are distributed so that they can be completed. They must then be collected and processed by the recipient ((i.e. the meaningful data must be extracted and captured from them). Finally, the forms must be filed away where they can be retrieved when needed. Each of these tasks consumes corporate resources. Document imaging and management software systems have addressed the issue of distribution and filing, but automatically capturing the data from an image, using a variety of recognition technologies supplemented by rules, is still evolving. This combination of document imaging and recognition technologies can provide major reductions in the cost of capture and indexing.

The Purpose of This Book

This book is designed to provide a basic understanding of the technologies employed and the expectations for forms processing systems. It discusses the complete relationship of a form to the business process — from the design of the form to improve its ability to be captured, through the capture process.

It looks at the types of forms that are currently being automated and the technologies being used to develop these systems to help the reader understand what is important to the would-be user and what is not.

It will also discuss future efforts to improve the automated processing of forms, and the impact of the Internet.

An appendix presents a number of successful case studies of users of forms processing systems.

Why Now?

Despite many individual examples of the capture of data from specialized forms being automated by using special fonts or optical marks, most forms processing and data entry from forms is still carried out manually within corporations. However, things are changing fast under the influence of

- higher speed processors allowing much improved recognition capabilities with better validation,

- high resolution graphics displays at the desktop, and

- high speed internal and external communications.

While document Imaging has been accepted for filing and storage by the mass market, leading edge companies are expanding these systems into a transaction environment to extract the data from and process the data from standard business forms electronically while capturing the image of the form for electronic management, storage and retrieval.

These systems all operate in a network environment with different operating systems (DOS, Windows, UNIX), although there is a strong trend towards 32 bit Windows based systems.

The systems are all based on using a variety of character and other recognition technologies to improve the accuracy of automated capture of information from forms. This is then supplemented with validations and rules to insure accuracy. Process based workflow distributes and controls the process across multiple PC's. Improvements in worldwide communications, including the Internet, are enabling users to keep the original paper locally, while lowering costs through shipping images to service bureaus and then to low cost off-shore locations for data entry from image.

Background

Billions of forms are used every day to control the transfer of information between individuals or companies. They pervade almost all aspects of our life. Much of the information from all these forms is looked at, acted upon, and captured into computer systems for manipulation and totaling.

Most forms processing and data entry from the forms, is carried out manually within corporations, often by clerical staff who have other responsibilities. It is extremely costly — we estimate that over $15bn. a year is currently spent by corporations in the industrialized world to capture information from forms. But the information we capture is driven by the programs that manipulate it — therefore it is often limited. Much information therefore stays on the original form — some estimates are as high as 60%. In addition to the ever-increasing volumes of data that we generate, we are also increasing the velocity of the information as networks and PC's enable the movement of a form image rapidly from one desktop to another.

What is happening?

Understandably, companies are looking to reduce this cost. But they are also recognizing that there is important marketing and competitive information remaining on many of these forms that could be usefully mined if the cost of collecting it is lowered.

Some solutions involve moving some of these forms to direct entry using the Internet or implementing 'electronic forms'. For example airlines who issue over 400,000 tickets a year have moved many tickets to electronic origination substantially reducing the need for data entry.

Other solutions use a mixture of: direct capture of information; electronic forms; and a combination of paper forms redesign combined with recognition technologies. Portable notebook PC's and communications, including the use of cellular phones and the Internet, is expanding direct capture to the places where the work is carried out or allowing customers to input their own data.

Despite these efforts the volume of paper forms still seems to be increasing, although the number of pre-printed forms is decreasing. While company created forms are increasing, forms processing solutions become a critical element in reducing cost.

This is being helped by

- standards, such as portable file formats, which allow forms based information to be displayed on any system within or outside the organization.
- recognition systems which can now automatically recognize the form type, identify, and capture the information into computer usable data..
- improvements and lower costs in worldwide communications, including the availability of ISDN and the Internet, enabling form's images or part images to be transmitted to lower cost off-shore locations, where key data entry or OCR repair from image can be carried out.
- improved validation rules
- better local area networks which can efficiently control the capture and movement of the forms around the organization.

CHAPTER

Defining A Form

Why are forms important and what is a form for?

The world we have created runs on the basis of using forms. Without forms our society would collapse. All our governmental systems rely on forms to capture and control the processing of information about its citizens, its infrastructure, its industries, and its transactions.

All businesses, whether they are in the financial market, the commercial market, the medical market, or the retail market, use forms to control, capture and transmit information. These forms are a basic tenet of every business, whether it be internal or external, from Accounts Payables - Zero Coupon Bonds.

Forms, on a personal level, govern and control our lives from birth to death. In a lifetime we receive and complete literally thousands of forms. Our introduction into the world is announced with a form - the birth certificate - and we are introduced to forms at elementary school with multiple choice questions. From there it gets progressively worse as we need to: get a driving license; file tax returns; open the first bank account; get loans or a mortgage; apply for credit cards; apply for insurance and make claims; get medical care, and complete warranty cards or surveys.

What happens to all these forms?

They are looked at, copies may be sent to various other interested people, information is extracted from them and is placed in databases or manipulated and totaled. Meanwhile the original forms are filed or destroyed.

Volume of Forms

The numbers of forms we use is staggering — the printing forms industry alone is worth $9bn a year. Some brief statistics will give an idea of the size of the market as all the following transactions required one or more pages of forms to be completed, read and processed. In each year in the early 1990's there were approximately:

- 28 million new life insurance applications in the US to be processed,
- 920 million packages shipped by express transportation,
- 6 million new mortgage loans processed
- 56 million new credit cards issued,
- 9 billion subscriptions for magazines
- 420 million passengers on the airlines

All these applications used forms — whether they be loan or insurance applications, waybills or tickets.

The different forms we use cover the letters of the alphabet: Accounting Changes, Bills of lading, Checks and Credit Card Statements, Debit Advises and Dividends, Expense Reports, Financial Statements, Guarantees, Health, Insurance and Invoices, Journal Entries, Loan Requests and Leases, Mortgages, Purchase Orders, Resumes, Statements, Shipping Documents and Surveys, Tax Returns, Telephone Bills, Tickets and Time cards, Utility Bills, Vehicle Registrations, Water Bills etc..

All these, and many others, are readily recognizable to the average person. Collectively we use thousands of them every day.

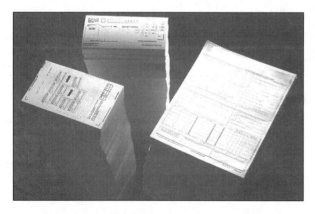

Figure 1.1 We use thousand of forms every year.

All these forms are different, but all have one thing in common - they seek to transfer information from one person or company to another in a way that is understandable to the average human.

It is that last issue *"understandable to the average human"* that causes the problems when we come to try to understand them in the computer and extract the data from them.

What Is A Form?

When someone talks about a form, most people instinctively think they know what is being talking about. In the generic context, a form usually consists of a series of lines and boxes with typed information and originating organization or company logo on top of the form. This type of form is a structured form.

A structured form can be categorized into three main areas of information

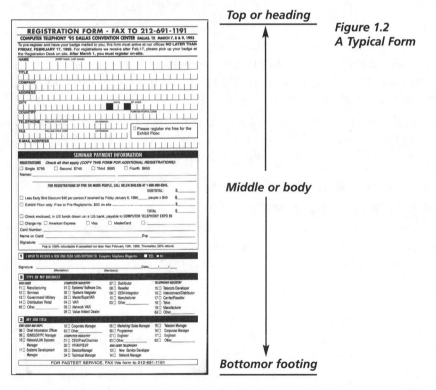

Figure 1.2
A Typical Form

These categories are:

- top or heading
- middle or body
- bottom or footing.

The top or heading part of a form contains originating company information, and areas in which overall information concerning the recipient company or person completing the form is located.

The center of the form, which can consist of more than one page, contains detailed information and totals, often delineated by lines or boxes. Some boxes contain explanatory text in them.

The bottom or footing of the form (which may be on the back of it) contains overall instructions and boilerplate or legal text.

A complete form set often consists of multi-part sheets in different colors to designate recipient or distribution. Each sheet in the set normally consists of lightweight paper separated by carbon paper or each sheet is coated with non-carbon coating.

Printed forms are either in single sets (known as 'cut forms') in which case the whole set is held together with glue at the top. Or they are continuous with glue and perforations at the sides suitable for a computer continuous tractor.

As high speed laser printers become common, more companies are printing the form layout at the same time as the content. In these cases the paper tends to be thicker and the form looks more homogeneous.

When Is A Form Not A Form? - The Unstructured Form.

The term "form" is used loosely to define a piece of paper that conveys structured information from one person or place to another. Everyone thinks they know what a form is and when we consider classics like invoices or statements they are obviously forms. A bank check is a form.

But some forms are more difficult to classify. Some forms are so loosely structured that they cannot be considered a form in this book, as they do not have characteristics that can be identified and extracted. An example of a loosely structured form could be a Resume. It has enough structure to identify dates and positions held, but is free-form in terms of exactly where the information is located, how much is contained in each section or even exactly what information is there.

However, a note from department A to department B on a slip of paper cannot be considered a form although the information contained on the slip is typical of a form's information.

Forms Must Be Understandable

Forms are used to transfer information from one person or company to another, so they must be inherently understandable by the recipient who may never have seen that particular form before. When they get to the recipient, actions must be taken:

- the form must be looked at and identified.
- the form must be processed.
- on the basis of the identification, the data must be entered or extracted or the information on the form may cause other forms to be generated.
- the form may be held up pending more information;
- it may be simply filed.

If the form is returned to the originating company, the new data must be extracted, and the form filed.

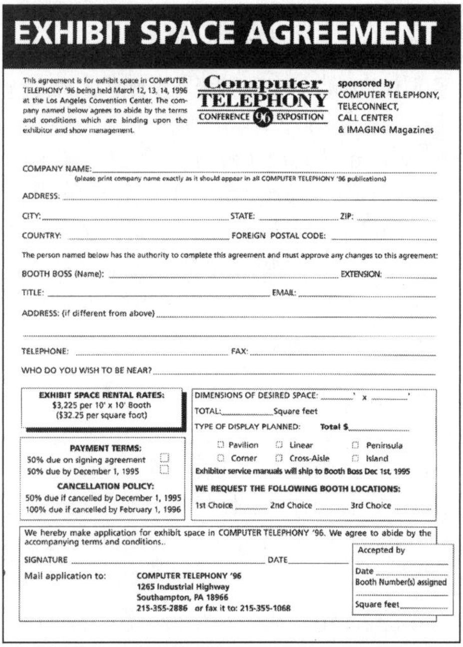
EXHIBIT SPACE AGREEMENT

This agreement is for exhibit space in COMPUTER TELEPHONY '96 being held March 12, 13, 14, 1996 at the Los Angeles Convention Center. The company named below agrees to abide by the terms and conditions which are binding upon the exhibitor and show management.

Computer TELEPHONY CONFERENCE 96 EXPOSITION

sponsored by COMPUTER TELEPHONY, TELECONNECT, CALL CENTER & IMAGING Magazines

COMPANY NAME:

(please print company name exactly as it should appear in all COMPUTER TELEPHONY '96 publications)

ADDRESS:

CITY: STATE: ZIP:

COUNTRY: FOREIGN POSTAL CODE:

The person named below has the authority to complete this agreement and must approve any changes to this agreement:

BOOTH BOSS (Name): EXTENSION:

TITLE: EMAIL:

ADDRESS: (if different from above)

TELEPHONE: FAX:

WHO DO YOU WISH TO BE NEAR?

EXHIBIT SPACE RENTAL RATES:
$3,225 per 10' x 10' Booth
($32.25 per square foot)

PAYMENT TERMS:
50% due on signing agreement ☐
50% due by December 1, 1995 ☐

CANCELLATION POLICY:
50% due if cancelled by December 1, 1995
100% due if cancelled by February 1, 1996

DIMENSIONS OF DESIRED SPACE: ' x '
TOTAL: Square feet
TYPE OF DISPLAY PLANNED: **Total $**
☐ Pavilion ☐ Linear ☐ Peninsula
☐ Corner ☐ Cross-Aisle ☐ Island
Exhibitor service manuals will ship to Booth Boss Dec 1st, 1995

WE REQUEST THE FOLLOWING BOOTH LOCATIONS:
1st Choice 2nd Choice 3rd Choice

We hereby make application for exhibit space in COMPUTER TELEPHONY '96. We agree to abide by the accompanying terms and conditions..

SIGNATURE DATE

Mail application to: **COMPUTER TELEPHONY '96**
1265 Industrial Highway
Southampton, PA 18966
215-355-2886 or fax it to: 215-355-1068

Accepted by
Date
Booth Number(s) assigned
Square feet

Figure 1.3
Seasonal Form Volumes peak close to show

Volumes Vary

Some forms types, such as shipping documents, may be received in small volumes regularly throughout the month with some variation at week's or month's end.

However much form information is very high volume and seasonal in nature - i.e. it is received at peak times such as the end of the quarter, end of year or around specific holidays or in connection with an event such as a trade show (see Fig. 1.3).

Seasonal data may be either required quickly or it may be possible to spread it over a longer period. For example order entry for a catalog company must be processed immediately, but warranty information can be entered in a more leisurely fashion.

Many companies will send much of this type of work to a dedicated Service Bureau who can even the load from different companies — but it costs a lot more to process seasonal high volumes quickly.

Environments and Usage Varies and Affects Automated Processing

Forms vary in how they are handled and the amount of care taken by the user. Some forms are used once in a clean environment. Internal surveys or school tests might fall into this category. The form is handed out manually in an office environment, completed by one person and returned for processing.Other forms such as shipping documents or auto repair bills have a more difficult procedure. Often the third or fourth copy is the one that must be worked with. They are moved from one place to another, handled by multiple people, often used in dirty or difficult environments and are not a primary document i.e. they are supplementary to a main transaction. The user has little or no interest in ensuring that the form stays as a quality document. These types of forms often consist of thin paper, they get dirty, they get torn and they are often difficult to read (see Fig. 1.6).

Who Processes Forms and What Does it Cost?

Forms are usually processed and the data captured by white-collar clerical staff, often as a part of another job such as analyzing the data or balancing the books. In many environments such as an accounting office, a member of staff will both process forms and create new forms to inform someone else to take an action.

Internally keyed data may be distributed to a data entry department if it is high volume and easily identified. There it is entered by a defined data entry staff. In this situation very high key entry rates are normal.

Normal business type forms are distributed to the department responsible for the data e.g the Accounts Payables department, and keyed by the regular clerical staff in an interactive fashion as they

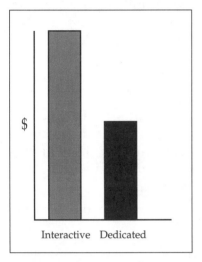

Figure 1.4
Cost of Interactive Vs.
Dedicated Data Entry
Interactive Key Entry averages
5 forms Per Day Over 1 Year
Dedicated Operators process
2-3 times as many.

are received.

This method of operation became particularly popular in the 1970's for accounting data entry in smaller and mid-sized companies, when interactive mini computers were used extensively to manage the accounting and control functions within offices. These systems were designed around a centralized database structure with text based terminals which could interact with the computer and perform on-line validations and retrievals of data that could be visually verified. It is a method of operation that has been expanded with the universal usage of PC's and networks. The cost of this data entry is largely concealed as the data is keyed as a small part of someone's overall job.

The problem with interactive type operations by part time key entry staff is that the operators are not terribly efficient. Their job titles do not reflect a key entry job and this is not the primary training — in addition the types of forms and data that must be captured vary tremendously.

Industry averages show that a dedicated key operator can key at 11,000 to 18,000 keystrokes per hour. However, inexperienced operators or those who only occasionally key data will generally achieve substantially less than 5,000 keystrokes per hour.

Consider that an average clerical worker who can type at 30 words/minute will actually be keying around 1,800 alpha/numeric keystrokes per hour.

In addition, clerical staff are paid substantially more than data entry staff. A data entry operator is paid around $9 an hour, according to department of labor statistics, whereas a typical clerical staff member may be paid $11, $12 or more per hour, depending on the location and the job title. In some cases you find accountants earning $20 or more per hour keying data! When overhead is added, which averages 2.3 times, the cost of interactive data entry is extremely high.

It is difficult to quantify exactly what this is costing us. Based on an analysis of the quantity of forms processed each year in a number of vertical industries we have concluded that very large sums of money are spent keying data. I estimate over $20bn a year worldwide is spent keying the data from the forms into the computers.

It is no wonder that many companies are targeting this area for automation.

Categorizing Types of Forms by Likelihood of Success in Automation

Forms can be categorized according to who creates them, who completes them, who uses them and what their purpose is. Each categorization is important in one way or another, as we will see. In

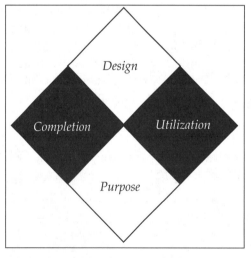

Figure 1.5
The four elements of forms processing

principle though, the more control over the creation and completion of the form that an organization has, the more chance that organization has of automating the process.

There are 4 elements involved in determining the chances of success in automating the processing of paper forms.

Design

The person who is responsible for creation of the form (i.e. design and printing) is important.

If they are also the beneficiary of the form, then they will be more inclined to develop a form that can be easily automated. For example: up to recently, banks controlled the printing of checks and they mostly consisted of bland backgrounds with black printing. With de-regulation, third party vendors started to print checks with all sorts of exotic colored backgrounds that can interfere with the data written by the user.

Completion

The person who completes the form is important

If you have an interest in the result of completing the form, you will be more likely to take some care and obey instructions. Let's take an extreme example where a pupil in grade school will pass an exam if they complete an optical mark form carefully with a number 2 pencil but fail it if they complete it sloppily or use a pen. The chances are that

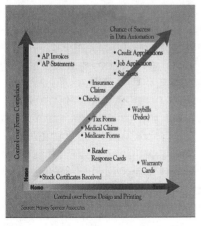

Figure 1.6
Success with forms processing depends on the amount of control you have over the design and completion of the form

they will complete it as requested. Rewards can vary and are usually not as extreme as passing or failing an exam. The US Postal Service attempts to reward a person filling in the 6+4 zip code by promising faster delivery. On bulk mail the Post Office reduces the cost if the sender barcodes it thus improving automation.

The opposite is a company requesting someone to complete an anonymous survey with no reward. This company had better design their form to be foolproof as the person completing it is only doing it as a favor.

Purpose

The purpose of the form is important

The purpose of the form controls the quantity and the accuracy of the data capture process. If a form contains background information for filing, then all that is required is a few fields, some of which may not even be on the form, to be captured for indexing retrieval purposes.

At the other extreme is a customer survey. In this case, all the data on the form must be accurately captured and then the form is no longer required.

Utilization

The company who uses the form is important

If the company who issues the form is the same company who uses the form, then the design of the form and processing can be optimized. For example: a survey may be distributed by a market research company who will receive the completed forms. They can place a barcode, which is easy to interpret, on the form with a reference number on it. When the form is returned, it is easy for the computer to read and classify the form. Likewise, a company sending out remittance advices with a bill can send an envelope with a box number on it to provide pre-sorting of the mail, and code the remittance advice with the account number. Even if only 10% of users return the remittance advice it reduces processing costs, and the incentive to use an envelope is the convenience and if it is pre-paid, postage cost. Banks have used this for years by encoding account number, check number, and routing codes in a specific font with magnetic readable ink on the bottom of the checks.

CHAPTER **2**

Defining & Evaluating Forms Processing

When does Forms Processing Make Sense?

Automated forms processing reduces the need for manual data entry from paper forms by converting the paper form into an image representation and applying recognition technologies. The value is two-fold

1 It reduces the expense of manually keying the data

2 It captures an image of the paper that can be distributed electronically and filed and retrieved almost instantaneously when necessary

But it costs an average of 4-8 cents to scan a page. At a loaded cost of $10 per hour for a key entry operator this means that and productivity of say 12,000 keystrokes/hr, this means that between 48 and 96 characters can be keyed for the cost of scanning. Using low cost off-shore labor (excluding transportation costs) you can probably multiply the number of keystrokes by a factor of 2-5.

There may also be a value in capturing the image depending on the application. If the original paper is required for audit, supplemental processing, or customer service then the cost of scanning it is absorbed by the attractions of having the image.

Data Mining Possibilities

There is much additional data often held on a form. Automated Forms Processing costs are fairly inelastic. Forms processing can increase the amount of data captured for minimal incremental cost.

Attachments and Supplemental Documents

Traditional forms processing has concentrated on capturing the data from the mainline form. For example in a Mortgage Processing application, the data that is to be captured is probably that on the front of the mortgage application form alone. There is much more information attached. Capturing the attachments at the same time as the master forms

- reduces document preparation;
- provides supplemental information that can be mined with recognition;
- and provides an electronic facsimile of the attachments that can be fed into a workflow or storage and retrieval system.

Escort Data

Escort data can consist of supporting documents but it is a term usually used when a form is linked electronically to another source. This can happen with a barcode or through textual links, but the escort data will normally consist of additional data that applies to the form where the linkage occurs.

The Main Steps of Forms Processing

Once you have decided to use forms processing you need to consider 5 steps:

- Forms Design
- Forms Data Creation
- Forms Data Capture
- Forms Distribution and Management
- Forms Storage

Forms Design

Forms design represents the original creation of a form. Traditionally this is a task for the art department who design the form in conjunction with MIS and the originating department.

The main point is that the average form is designed to be

- easy for humans to understand,
- easy to extract information from, and

- easy to process by the average human whether or not he or she has previously seen it.

As with all types of publishing, the typical approach to forms design has changed, and continues to change, with the increased use of desktop publishing which enables the user to quickly draw up a form and print it.

Alternatively electronic forms design tools are allowing a user to scan an existing paper form and convert it from bit mapped "raster" data into "vectored'" computer data and fonts which allows easy modifications to occur.

Also there are some forms design systems which assist in creating forms that will be both user and computer friendly.

Forms Data Creation

Forms are used to transfer information. Depending on the application the information or data on a form is created through different methods. These may vary from individuals equipped with a blunt pencil operating in cramped and dirty conditions to computer generated laser printed forms and data - straight, nicely located, and clean. As with design, the data on the form is designed to be easy to understand by humans.

Forms are completed as follows in order of potential of recognizing the data starting with the most difficult:

Hand Written Forms

Hand written forms are usually used for consumer applications. These types of forms are often supplemented with check-off boxes that are easy to use and which are fairly simple for a computer to interpret.

Single Set Typewritten Forms.

Typewritten forms are becoming more rare as computers continue to be used in small businesses and the home. These forms are placed in a impact typewriter or sometimes a dot matrix printer for completion. They sometimes consist of multiple parts.

The data on these forms can be difficult to read on the front copy because

- the layout boxes are often cramped in their design and typewriters or printers can be difficult to line up precisely.

- They can be even more troublesome on the copies as:-
- the impact printers are not strong enough to create a readable copy
- carbon or encapsulated ink can leave extraneous marks on the copies

Multi-part Continuous Forms

Probably three-quarters of the forms that are traditionally used in business are multi-part forms that are completed by a computer. Typically, the form is placed as part of a continuous fan-fold set, lined up with a check pattern of some sort and then left to run. As the printer's tractors pull the paper through the transport, the paper adjusts somewhat, causing the printed type to cross over some of the boxes or skew a little. It is easy to understand for humans, but causes difficulties for computers. As with cut sets, these types of forms are also often carbonless - that is the sheets are coated with tiny encapsulated bubbles of ink (known as NCR paper), which then cause a blue image of the typing to occur on the copies. This not only gets faint on multiple copies, but if a multi-part set is placed in the mail, the bubbles can burst along the folds, causing blue lines to occur in the middle of the form.

Carbon copies are often used in multi part forms that are placed in a typewriter or sheet fed desktop dot matrix computer printer. While these are better than the NCR papers used in high speed printers they exhibit some some of the same problems and the carbons in the second and subsequent sheets can cause extraneous lines and marks.

The front copy of both these three types of forms is subject to image degradation if the ribbon is not monitored.

Laser Printed Forms

Forms printed on laser printers are being used increasingly instead of multi-part forms. They are simple to deal with if the toner is fresh. As the form is normally an electronic form, which is created on the printer at the same time as data, everything is well lined up and the data does not impact on the form. However unless the laser is a color laser, the background form is the same color as the data which can cause problems with removal with some systems.

Forms Data Capture

Forms data capture represents the extraction of required user data from the form to be processed. Only certain amounts of the overall data contained on a form are ever required. This may vary from one or two fields to all of the variable information printed on the form. A typical purchase order may only require that limited fields be captured by the purchasing department — for example a supplier number and details of the items that are to be purchased. But a warranty card or a survey will require all the information that was entered to be captured. In some cases as in when processing medical claims, information is displayed in one format but must be translated into a code by a trained key operator while keying it.

Traditionally clerical staff are trained to be able to recognize what information is required, where it is located, and what to enter into the computer. They are also trained to ignore extraneous information that may be required for reference purposes, but which is not required for processing (e.g. signatures) and ignore information that is instructional (e.g. forms completion instructions). This training is partially generic i.e. if a box is labeled "Invoice Number" and the computer's entry screen is requesting the Invoice Number, then it's obvious to an operator who understands English what to enter. But some data needs industry specific or company specific knowledge to identify the required fields or the conversion of the printed or written data into defined codes.

Forms Distribution and Management

Completed forms must be distributed and processing by the people responsible for it. Typically a form, which may consist of one or more pages and may or may not have attachments justifying a field or as proof, is distributed to the people responsible for making decisions regarding the processing of it.

Some typical examples include:

- Claims adjusters receiving insurance claims. If the claim does not satisfy certain criteria or exceeds certain parameters, it may get placed in a different queue, or it may be handed on to a supervisor or another more specialized staff member for special handling.

- Check payments which if over a certain value must be authorized by a supervisor.

- Capital expenditure requests have to be approved by different levels of management depending on the amount, the purpose of the expenditure and/or type of capital, and possibly the dates in which the expenditure is to be incurred.

- Tax returns which do not fit certain audit control checks. There may also be randomly selected ones get routed to an investigator based on the type of business and location of the business or individual that sent in the return.

Figure 2.1
The Flow of a Form Through an Organization

Distribution Method

Traditionally a company's mailroom will receive and sort the mail into categories, and departments based on the address on the envelope or based on the content.

Depending on the specific work process, invoices and other financial transaction, forms may be extracted and possibly batched for entry before sending on to the payables or other responsible department.

The individual envelopes with contents destined for specific departments are then dropped into departmental bags or bins, loaded onto mail trucks or passed to an internal mail delivery person and dropped off in the department responsible. Someone in that department is then responsible for opening the mail and sorting it into different categories, depending on type of mail and the action required.

In larger companies the payments process has been automated for some time using post office box numbers. Usually located within a postal area close to the sending person, post office boxes allow payments and forms containing remittance advice information to be received with minimal distribution delay. Information can be pre-sorted using department numbers or different PO boxes.

In most systems the payment and remittance advice is opened by a mailroom at the Post Office box location. It is then captured using automated remittance advice systems by the company's bank at the post office box location, and transferred electronically to centralized collection accounts. Any exception item forms needing specialized attention such as those with comments or change of address can be forwarded to the company after the payment is extracted. These systems, which are essentially a type of forms processing system, are rapidly moving to image based processing too. This is because you can process more forms and the exception item forms can be processed more efficiently using recognition technologies supplemented by key-from-image — effectively eliminating paper.

Other types of forms distribution and handling is not so automated. Now the forms are beginning to be distributed on an ad-hoc basis through E-Mail and the Internet is being used to input forms data directly— effectively eliminating paper.

Forms Storage

This represents the storage of the forms after initial processing, or the permanent storage for archive purposes. In a traditional paper based system, documents are photocopied and filed in multiple different places under many different headings. Each department keeps one or more copies as they process them. Eventually locally filed copies are destroyed and one copy ends up in archives. Often it is microfilmed to reduce storage space and then it usually becomes accessible only by a records manager.

Document management systems eliminate the need to copy the information and file it separately at a department or individual level. They also reduce the need to microfilm by providing an archival quality record of the transaction. They achieve this through taking a scanned bitmap representation of the form that can be easily transmitted across a network and displayed or printed as needed. It is typically stored temporarily on magnetic disk while people work on it, and then placed onto optical disks, CD or DVD for long term archival storage.

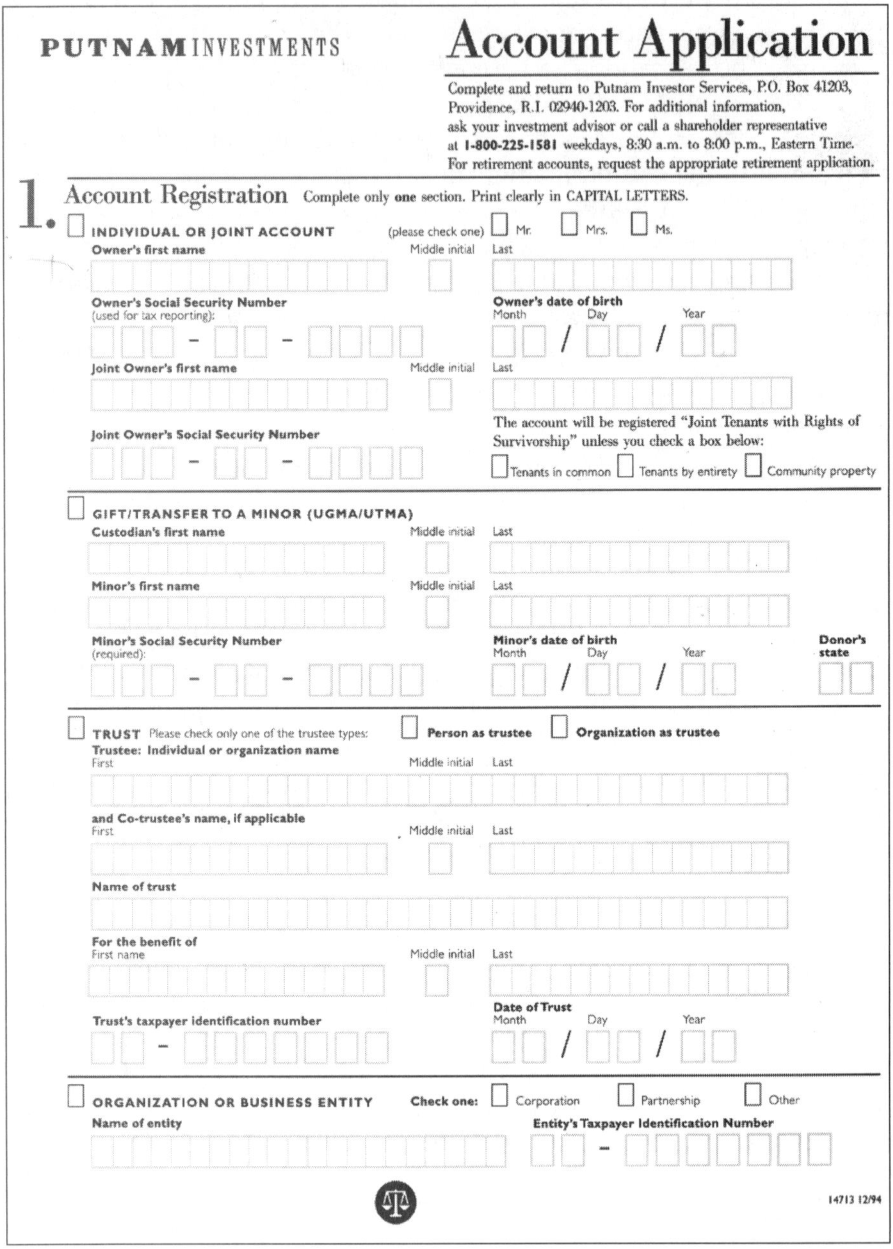

Forms Design - A Form Defines the Company

PUTNAMINVESTMENTS

Account Application

Complete and return to Putnam Investor Services, P.O. Box 41203, Providence, R.I. 02940-1203. For additional information, ask your investment advisor or call a shareholder representative at 1-800-225-1581 weekdays, 8:30 a.m. to 8:00 p.m., Eastern Time. For retirement accounts, request the appropriate retirement application.

1. Account Registration
Complete only **one** section. Print clearly in CAPITAL LETTERS.

☐ **INDIVIDUAL OR JOINT ACCOUNT** (please check one) ☐ Mr. ☐ Mrs. ☐ Ms.

Owner's first name | Middle initial | Last

Owner's Social Security Number (used for tax reporting):

Owner's date of birth | Month | Day | Year

Joint Owner's first name | Middle initial | Last

Joint Owner's Social Security Number

The account will be registered "Joint Tenants with Rights of Survivorship" unless you check a box below:

☐ Tenants in common ☐ Tenants by entirety ☐ Community property

☐ **GIFT/TRANSFER TO A MINOR (UGMA/UTMA)**

Custodian's first name | Middle initial | Last

Minor's first name | Middle initial | Last

Minor's Social Security Number (required):

Minor's date of birth | Month | Day | Year | Donor's state

☐ **TRUST** Please check only one of the trustee types: ☐ **Person as trustee** ☐ **Organization as trustee**

Trustee: Individual or organization name
First | Middle initial | Last

and Co-trustee's name, if applicable
First | Middle initial | Last

Name of trust

For the benefit of
First name | Middle initial | Last

Trust's taxpayer identification number

Date of Trust | Month | Day | Year

☐ **ORGANIZATION OR BUSINESS ENTITY** Check one: ☐ Corporation ☐ Partnership ☐ Other

Name of entity

Entity's Taxpayer Identification Number

14713 12/94

Figure 2.2
Example of a well laid out form

Forms Design is key to success in automating the forms process. A form is often the main item that projects your company's image to the outside world. It is an advertising and promotion item. How your form looks is going to impact on a supplier's or customer's perception of the company as much as the letterhead or advertisements. It therefore makes sense to spend some time thinking about the design of your forms. However, sadly many companies do not spend enough time thinking about design.

It is preferable not to delegate forms design to an external organization unless you insure many reviews.

When redesigning a form or designing a new form make sure that everybody who is involved in creating or using the information is consulted. If it does not work for someone, then that person will resist using it and quite likely find another way to transfer the information — ruining the best-laid plans. Departments involved should include:

- All user departments (including those who create the form and all those to whom it is distributed);
- MIS
- Graphics Design
- Management
- possibly Human Factors

Apart from the image of the company a form projects, its design is important for many things, including

- ease of use by the person completing it,
- ease of use by the person looking at it,
- ease of distribution and
- security of information

All these issues need to be considered when designing a form. As such many people are involved in the sign off and the smallest item can often hold up a form for unexpected reasons.

There is now a trend towards color scanning and capturing the initial image in color. This makes life a lot easier as the designer does not have to conform to specific colors, the form layout can be bolder and rescanning is minimized. However the software has to be able to deal with a color image.

The Corporate Image:

In a recent discussion on forms design with a client who was in the fashion business uncovered that

image was all important in all that corporation's relationships with suppliers, customers and other any other outside people. Therefore any forms had to conform to the corporate image in color, texture of paper, fonts and aesthetics. Sometimes these needs can conflict with the design needs of a form that can be easily automated. Ensure that you get the artistic design department working closely with the MIS department in all forms design.

Ease of use

A well-designed form is logically laid out (See fig. 2.2). It is easy to understand and it is obvious who is responsible for what part of it. It is easy to find the information that one particular user needs. It is easy for the person completing it to find and complete the areas he needs - there is no duplicate information and following from one form or one side to another is minimized. If it is a two sided form it is logical in what each side contains.

The well designed form is structured so that both the person completing it and the user(s) can follow the order of the information without needing to hunt for it. It is clear what happens to the form or part of the form next.

Ease of Automation

Unfortunately ease of automation factors sometimes conflict with human factors and PR issues. Automation is easier when using simple black text on white backgrounds, light pastel drop-out colors for the form layout, simple fonts, large spaces, barcodes and check marks and specific identification and/or location marks. Look at figure 2.2 which represents a good design for a form, but some systems may have problems processing it due to the lack of registrations marks. Many of these ideas conflict with the ideas of the designers and those who matter. They want to use BOLD colors with lots of contrast, fancy backgrounds, exotic fonts etc. Personally, I believe that automation should follow other issues not dictate them - and with the move towards color images, it becomes unnecessary to worry about drop-out forms or specific colors. However as most recognition is based on black and white patterns, color introduces problems in applying automatic recognition.

Security

Because forms are often used to satisfy many different criteria and pass different information to different departments within a corporation, some copies of forms contain shadings on second and subsequent copies. In addition some forms, such as stock certificates, are designed specifically so that they may not be photocopied. These needs sometimes conflict with automation and since compression is based on run lengths of a specific color, shade security scrolls can increase image file sizes dramatically, and the security scrolls can interfere with character recognition.

CHAPTER 3

Automating the Process

The first thing people who do not know how computers work, usually say about computer recognition is "I can read the form so why do I need complex software?". Once they have understood the potential then they think recognition will identify everything. Unfortunately, both positions are equally fallacious.

The main problem is that computers can't read images, they can only display or print them. Computers use binary codes, collectively known as ASCII, to represent characters. These characters are normally displayed using a notation known as Hexadecimal, since the numbers 0-9 plus the letters A through F represent the numbers 0 to 15. As an example of a computer's binary reqpresentation using Hexadecimal: 48 41 52 56 45 59 20 53 50 45 4E 43 45 52 is the notation used to display the ASCII characters for HARVEY SPENCER. ASCII is completely meaningless to the average human. So programs have to be made to convert human readable characters to computer readable (i.e. ASCII). These programs are known as OCR for Optical Character Recognition or ICR for Intelligent Character Recognition.

As forms were invented before computers, they are mostly designed to be easily understood by humans. It is easy for a person to identify where a supplier's name and address is located, it is usually pre-printed at the top of the form. And your account reference number is simple to find as it usually resides in a box labeled "YOUR ACCOUNT NUMBER". But the form and the information is not as easily understandable by a computer. To be useful in transacting today's business, the form must be routed to the responsible department, read, acted on and the information on it transferred from the form into the company's computer system. There it can be manipulated, added to other information, and accumulated into statistics.

When a form is scanned, the scanner makes an image, normally—with black and white dots - typically 200 or 300 dots-per-inch (dpi). Black is represented with a binary 1, white with a 0. When the form is displayed, it looks just like the original. Suppose you want to edit the document - fix a misspelling or add some punctuation. It's difficult as you must use a PAINT program and the computer cannot move the other characters along.

OCR, ICR and other recognition technologies solve this problem - they convert the patterns of dots to computer usable data - ASCII. Then, the data fields can be placed into a retrieval database or processed.

Routing, control and action on the forms is being developed through document imaging systems and workflow software. Capture and transfer of the information is being made more efficient through recognition technologies and key from image.

Image Capture Systems

Image capture systems have been developed to improve the capturing of images for processing and storing for future retrieval. In the simplest terms capturing the image requires scanning it. Then once it is scanned it is available electronically, but in order to manage it and store it for future retrieval, the user needs information concerning the type of data and some retrieval indexes. The first imaging and document management systems scanned each document (which could consist of one or more pages), displayed the scanned images with an electronic form that enabled the user to verify readability of the document and key the indexes. This was a very inefficient method of operation and it resulted in systems that had scanners installed capable of scanning a page in one or two seconds taking 20 or more seconds to capture the indexed page.

To solve this, a variety of vendors developed batch-scanning systems. These systems were designed to maximize the scanners through placing coded document separators between the documents. These coded separators (which usually use barcodes or easy to read text that can be converted with optical character recognition) then provided a basic index. The document preparation time increased a little, and the user had to scan more pages and potentially take out the separators after scanning, but the overall impact was less capture time and substantially less labor.

But these systems limited the retrieval indexes, so they were supplemented by key indexing usually implemented on the same basis as the manual approach - i.e. look at the image and key it. The difference is that in order to ensure that the scanners did not slow down, this key entry had to be carried out on distributed workstations.

This in turn meant that any images that were difficult to read had to be identified, the document rejected and the information concerning passed back. The original paper then had to be found, a manual rescan initiated and the new image re-inserted into the document. This caused disruption in the key entry process, so vendors started to implement automatic image clean-up and automated image quality assurance. The same processes that the forms processing vendors had implemented.

As a result there is an overlap between document capture systems and forms processing systems. Each are designed to take a batch of documents, convert them to electronic images as quickly as possible, ensure good quality images and capture information from them.

However there is a difference. Image capture systems usually carry forward indexes from one document to another if more than one document exists in a folder. Forms processing systems look at each document as an individual and unique document.

The process of capturing the documents and their associated data is managed by process workflow software and each vendor has developed some proprietary workflow usually based on an underlying database to manage this transaction flow.

Figure 3.1
a stylized example of automated data capture / forms processing processes

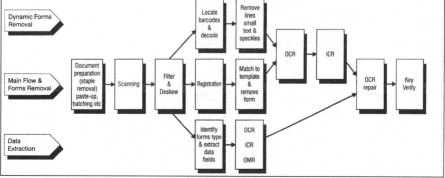

WORKFLOW

It is useful to review the term workflow and categorize it. Workflow is a term which is used to define the process of routing and controlling the flow of documents electronically through an organization.

Unfortunately it has become a catch-all phrase which is frequently used to encompass any form of image or business document distribution.

In the case of automated forms processing, Workflow can be categorized into three different areas:

- Ad-hoc workflow
- Pre-defined workflow
- Knowledge based workflow

These areas are not exclusive, so they can be intermingled to provide a solution. For example consider the following document flow:

- A payment is received, all checks and remittance advices are scanned and the checks are physically out-sorted and batched for deposit to the bank.
- Images of checks without remittance advices are sent to specific people in the receivables department to match up with appropriate invoices.
- The remaining remittance advices are sent to an automated reconciliation system - those that do not reconcile, are routed to clerical staff.

This simple example can be carried out with a pre-defined process workflow. But then, a particular payment may not fit any category and may have characteristics which require it to be "manually" processed. In this case, the payment will be placed into an ad-hoc workflow queue and managed separately.

Ad-Hoc Workflow

This typically utilizes electronic mail to distribute data on a "push" basis. That is, someone wants someone else to receive a form or other information so that person sends it to them for information completion, or some other action. In some cases they may attach a note to it or some other document that is relevant.

Predefined or Process Workflow

Predefined workflow utilizes a set of specific rules to control the distribution of a document. A simple example is FAX routing. Based on an OCR interpretation of the sending company, the FAX is automatically placed in a specific department's or individuals mailbox queue.

Predefined workflow is also used in front-end capture systems in forms processing as follows:

- One or more stations are used to scan the forms
- They are routed to Quality Control stations for viewing.
- Successfully scanned images are sent to pre-process stations, while those needing re-scanning are sent for out-sorting and re-scanning.
- The forms are identified and matched to templates
- Image zones are extracted based on the type of recognition required
- The images are sent to appropriate recognition engines
- The results are verified against tables and reasonableness checks
- Incorrectly converted images are sent to post-recognition repair
- Those requiring 100% accuracy are sent to verification operator stations.

The workstations in this scenario can be configured, and reconfigured, as necessary to perform different tasks.

Knowledge Based Process Workflow

Knowledge based distribution utilizes knowledge of a process to define its routing. These processes are defined as rules which may be based on

- types of documents,
- time scales,
- amounts involved etc.

Effectively Knowledge Based Distribution attempts to mimic the distribution decisions that humans make for information in the office. It delivers electronic images of the mail or document(s) to predefined people and keeps track of building up files that must contain specific items before being processed.

It tracks the process of documents as they flow through an organization, checking for additional documents that must be processed , checked or attached before a document can be released, tracking time scales and costs to process the documents.

Most workflow in forms processing is process based workflow. In simple terms this means that the forms are batched, scanned as batches onto a main server. They are then pulled down one by one, processed and the data exported (see fig. 3.1).

Internet Makes Process More Efficient

Using the Internet it has become possible to distribute forms processing workflow in a more cost effective manner while allowing the organization who needs the original media to maintain control ove it. The paper forms can be scanned at the distributed location where the paper may remain and the images managed for customer service etc.. The images of the forms can then be sent to a centralized IT operation or Service Bureau for processing and recognition. Finally they can be sent off-shore for key entry repair and verification.

CHAPTER 4

The Role Of Paper
In an Electronic Age

Low cost portable computers, lowered communications costs, universal access over the Internet and document imaging are changing the traditional methods of using paper based transactions to manage commerce and society.

Changes are occurring in three ways:

- Direct Capture.
- Data Entry by the originating organization.
- Recognition of paper forms with extraction of the data automatically supplemented by improved manual processes.

Direct Capture

Some industries are moving to direct capture. For example, the insurance salesman still comes to your home. But unlike previously, he no longer arrives with examples on paper, a pad of forms and a calculator. He now has a notebook computer and while explaining various plans to you and showing graphical representations of your options, he can capture your data directly. Likewise Claims Adjusters can automatically capture pictures of an accident, the location from a GPS positioning satellite, and enter relevant data directly into a PC at the accident site. It is then uploaded when the adjuster returns to the office.

When you apply for a credit card or an auto lease, credit information is extracted directly from one of the credit reporting agencies using your social security number. The agency utilizes the social security number to extract and download all credit details that can be automatically incorporated into the loan request. This reduces the amount of information that must be entered or the number of forms that must be completed. It also improves customer service and probably results in more sales as the credit approval can be almost instantaneous — before the customer changes his mind.

Ninety per cent of retail mail orders are now phoned in through toll free 800 numbers and entered directly by an order entry clerk at the other end of the phone.

Surveys are now being delivered on floppy disk, via on-line services especially the Internet or through tele-marketing. You answer the questions and send back the disk, or increasingly you answer the questions directly while connected through a modem. Alternatively, where questions may be more subjective or need explanation, a questioner reads the questions and keys the answers while you are on the phone. Likewise more and more warranties and product registrations are beginning to be entered and captured electronically.

Capture by the Originating Organization

We are in the beginning stages of a true information delivery revolution. As on-line services become more common, as delivery networks increase, as telephones get smarter and TV becomes interactive, information captured on forms will increasingly be captured on-line, reducing the number of paper based forms that must be completed. Surveys can be answered with a hand-held computer. Credit cards and loans can be applied for, and bills, paid through a PC and modem. Instead of filling out a form, on-line capture by the user when applying in person for a service will become common. One person will be connected with multiple different people on demand assisting them as necessary through the telephone or Internet voice.

As the new generation of smart devices gets installed one can envisage a future where most consumer or business devices will report their presence electronically removing the need to enter warranties.

Electronic Forms

Electronic forms creation and control systems are being developed with interfaces to databases to allow some types of forms to be created, managed and delivered electronically using networks or communications. Electronic forms deliver ASCII data, so they allow computers to easily extract and manage the information contained on them. For internal usage, electronic forms provide major benefits.

They are effectively a form of structured E-Mail that is easy for the computer to read

The information can be better controlled due to the structure inherent.

An expense form filed electronically can easily total up amounts and even produce "warning" messages if a user is exceeding corporate daily limits. Because it is electronic it is easy to identify relevant fields and set up and control the predefined or process workflow.

Externally created forms are a different issue.

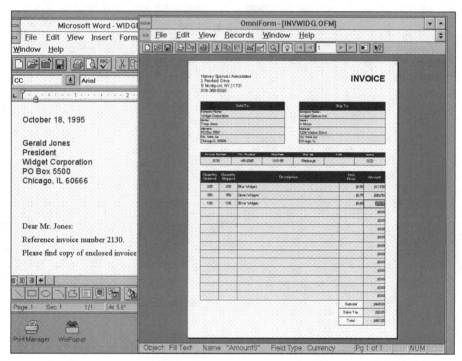

Figure 4.1 Example of an Electronic Form.
(Courtesy CAERE Corporation)

EDI was the first attempt to utilize electronic forms data that could be understood and directly inserted into the computer. But EDI with some exceptions particularly in the transportation industry has not been that successful. In some cases such as Medical Claims, the formats have been so variable that developing software to read them has been difficult.

However the Internet is changing things and driving standards. Increasingly users are being asked to enter data through the Internet. The first implementations of this are HTML forms that are 'submitted' to the Internet server who runs a CGI script to validate the fields, it then returns an answer. The second generations have used Java Script to validate the fields or downloaded Active-X or Java applets. The third generations will leverage from XML (Extensible Mark-up Language).

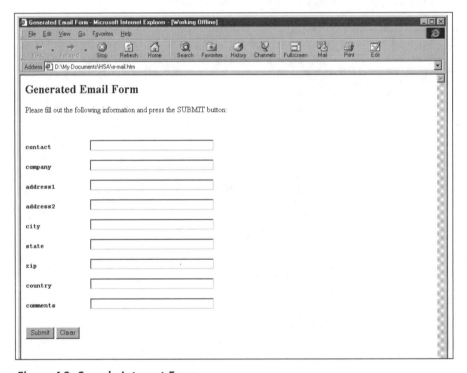

Figure 4.2 Sample Internet Form

One type of form capability has been developed by Adobe based on their PDF (Portable Document Format) formatted documents, which is a generic format that can reproduce the form visually on any PC. PDF was supplemented with the ability to create forms for data entry, validate fields based on Java script and secure signatures or data with 3rd party products.

Electronic forms over the Internet introduce a new element as they can carry other supporting information or messages such as voice or video, or connect to other locations for explanation.

However, whether a form is generated electronically or on paper, there is still a need to capture accurate data and the rules developed still apply.

There Will Still be a Need to Use Paper Forms

Off-setting direct entry into a form and entry by the originating company, will be the need to capture more information from more diverse sources. We always need more information and the computer and telecommunications industries continue to generate more and more information that must be captured, managed, and manipulated. Much of this is in a paper format and will continue that way for a substantial period. Many forms contain interesting information, which is not being captured currently since the cost of capturing the data outweighs its benefits. As document imaging systems become commonplace, the data from these forms will be used to provide additional management information.

In document imaging systems these types of forms are being scanned for management, indexing and retrieval purposes. Capturing the information contained on these forms — information that is not currently entered — will provide major benefits to users.

One example is the police report filled in at the site of an accident - a small amount of information is contained on these forms which could be useful in creating statistics for the insurance company handling the claim.

Another example could be doctor's medical claims that identify the times and hours of the work undertaken. Some unscrupulous doctors might be claiming that they are working in two places at once, but the insurance company doesn't currently know this.

A third example might be attachments to a loan request listing securities held indicating interests.

These types of information are becoming increasingly required to improve business processes or market more effectively. They impossible to find out at present, but by interpreting already scanned forms, it can become available at very little incremental cost.

In the case of forms capture, the scanned image of the form is designed to provide the capability to apply recognition to the form to capture the data from it and reduce key entry. The data on the form must be captured 100% accurately. It is rarely acceptable to capture incorrect characters. For this reason edit checking and in some cases blind re-keying is necessary.

The paper forms industry is reporting a reduction in sales of paper forms due to the advent of electronic forms but the overall usage of paper continues to grow. This has been mainly due to the usage of laser and other non-impact printers which allow users to create their own paper forms, so the number of forms used continues to grow but our total paper usage increases. The use of electronic forms will reduce the total number of paper forms to be processed, particularly those that are used internally within a company at present. Since the phenomenal growth of the Internet we have seen a major increase in the use of electronic forms. Clearly this trend will continue reducing paper forms but never eliminating them.

Document imaging systems are designed to allow forms and other business documents to be converted to an image - a bitmap representation that can be transmitted, stored and reproduced on a screen or printer. The key to document imaging systems is to be able to find the documents. This is carried out through a series of indexes that point, via a database, to the image file. Normally a document, which may consist of one or more pages, will be connected to a number of different retrieval indexes. A chronological file held for all documents can be simulated by indexing the documents by a date and time of receipt. An invoice will also have a supplier name, an invoice number, and possibly other indexes. In these types of applications, incorrect characters may lead to problems in finding the document, but some errors can be forgiven if there is another search method available. For example the first search may be by invoice number but if this does not turn up the document, then the user can use the vendor name and approximate date to find the document he needs. Alternatively if the retrieval engine is sufficiently clever then the user may be able to find the relevant document without a correct index.

Mixed Forms

Forms can contain a mixture of elements. For example a form can contain a barcode or coding that connects it directly to an electronic location on the internet. This capability makes the electronic repository supplemental to the paper and has the ability to add to the paper with dynamic media. content.

CHAPTER 5

Categorizing the Types of Form

Forms can be classified into various categories as below. Each categorization affects the automation methods that must be used and the chances of success in automating the capture and processing of those forms.

Data Types - Confined and Free-form Forms

Confined forms consist of those types of forms where the data is specifically located in defined and labeled boxes. Most items that we initially think of as "forms" such as credit applications, invoices, waybills, tickets, registrations etc. fit into this type of category.

Confined forms can normally be recognized through some form of mark or logo which is unique to that form. Provided that the computer has seen the form before and been trained to recognize it - the data can be found and extracted. Unlike humans though, forms which have not been "pre-registered" into the system - i.e. unknown forms, cannot be recognized and used successfully yet (see discussion on 'line removal' later).

Another type of form is a free-form form. Free-form forms contain data that is being transferred from one person to another. These can be considered to be forms in as far as they are captured and processed like any other forms and are often a part of an overall forms processing stack of information. The big difference is that the information is contained in a free-form format. Examples are resumes or letters. Usually most of the information contained on a free-form form is required to be captured, although some form of post-processing parsing is often required to remove unneeded content, or to code information about the forms or the data.

Single vs. Multiple Pages

Although many forms consist of multi-part sets, most of the forms we use consist of single pages which are single sided. That is, one side of a one-page form contains all the required information.

But some forms consist of multiple pages that may be two, three or more pages long. For example tax returns or consumer surveys. In this regard, a multi-page form may consist of a two-sided single page form. Although a double sided form consists of a single sheet of paper it consists of two separate images, after it is scanned. These may be captured as separate files or as a multi-page tiff which consists of two images combined into one file. Either way the system must deal with two separate images.

From a system standpoint whether it is a single page form or a multi page form makes a lot of difference. Single page forms can be batched and controlled in some defined quantity (normally people use batches of 100 sheets). The software knows that each page must be dealt with as a separate form. As pages are scanned, the software keeps a count and if the number of pages counted falls below the total, then the system knows there must have been a double feed and sends the batch to a repair queue.

Multi page forms or mixed multi and single paged forms cause more difficulties. If the forms vary for a given process, then either the pre-batching operator or the software must identify and separate each form based on some type of differentiating header pattern information. If, however, all the forms are uniformly three pages long then it is easier to recognize the start of each new form.

The convergence of document capture with forms processing means that multi-page stacks are much more prevalent but data may only need to be captured from the first page for indexing and data entry purposes. For example this occurs when there is a main form supported by attachments.

Paper Types

Paper varies dramatically in weight and weave. Onion skin paper weighs 9lbs and anyone feeding that into a photocopier knows the risks they take. Most business forms utilize lightweight paper of around 11lb stock where multiple copies are involved to reduce the cost of the paper and often, more importantly, the cost of mailing.

From a scanning viewpoint, this type of paper is a problem as:

- it can be easily torn when it is removed from the envelope or when it is separated and it can be difficult to feed through the rollers in a scanner's autofeeder.
- it can jam in the transport
- the weave of the paper tends to be loose which causes dust on the perforations as the sheets are separated or if continuous the sheets are burst and the holes removed. The dust causes rollers to become slick, which prevents them from feeding the paper effectively through the scanner's transport.

Sometimes an item is stapled to the back or the front of the form and when the staple is removed, a hole can be made in the paper. This can catch on the separator rollers and can cause jams. The hole also may obscure some important information contained on the form - but a typical scanner with a white background will interpret the hole as white space that may not be picked up by Quality Control.

Normal copier paper is around 18lb weight and laser paper 20-24 lb weight. Most autofeeders can reliably feed 16-25lb weight paper, although transitions from one to the other will cause double feeds on occasion. This particularly occurs when the transition is from thick to thin, where the rollers open to accommodate the thicker paper and cannot close down quickly enough to stop the next sheet from feeding.

Copies:

Many forms contain multi-part sheets separated by carbon coated sheets or are coated with pressure sensitive ink (so called NCR paper) - normally blue. They are usually held loosely together with glue. In some rare cases, the carbon can be painted directly onto the back of the form as in Airline Tickets. There are a number of problems caused by copies:

If carbon is maintained as separate sheets they can cause:

- black lines to occur where the folds were;
- smudged characters which are diffcult to read or bleed together
- other extraneous marks which interfere with legibility.

The blue image created by No-Carbon can be very faint on the fourth or fifth copy and suffer from some of the same problems as carbon based forms.

Any forms with carbon painted on the back can lose a little bit of that carbon onto the rollers of a scanner every time a page is fed. This builds up and eventually causes the rollers to stop feeding the paper properly and, in some cases, glues up the mechanism - which can be costly.

Some forms have been photocopied. While photocopying can improve legibility somewhat for people, it generally causes more problems than it solves for scanning and recognition. The main problems include skew and loss of information. The image can get skewed on the paper when photocopying smaller papers on an autofeeder and also when using a flatbed. Loss of information occurs when the image and any extraneous marks get "set". As a result, algorithms which work on the basis of using grayscale shadings of information that are not even visible to the human eye will not work properly to recover the image, enhance the contrast, or remove backgrounds.

Colors.

Forms are often identified by color and sometimes the color of the ink that defines the areas to be completed changes on the form, or the color of the actual information on the form changes. For example the airline ticket uses red carbon which creates red text when it's issued.

More normal though is for forms developers to change the background, especially for multi-part forms i.e. the blue copy goes in the file, the white copy to the customer, the pink copy to the shipper etc. While most pastel colors do not affect image quality, some of these colors can cause image problems. The easiest form to capture is the top copy of a white form printed in black using a laser printer. If this is not possible, blue ink on yellow works well. Beware of pink and in particular red print on a dark pink background.

Red is a problem as most scanners (i.e. all those that use green fluorescent lamps) are blind to red so it translates into black on a bitonal image. I've seen forms with red printed on black - this is disastrous for the average scanner. A particular offender, which interestingly enough was designed to enable automatic capture of information, is the HCFA -1500 medical claim form which is used to capture claims information. This form is normally red and information is printed in black. On older scanners, before the advent of fluorescent bulbs, tungsten lights were used to provide the high level of illumination needed by the CCD - these lights reflect red - hence the red forms which drop out.

To solve the problems of these forms, most black and white (bitonal) scanner manufacturers now offer a red-lamp version or kit which consists of red fluorescent or LED bulbs that eliminate all red colors from the image. Some scanners provide two bulb sets that are switchable using the control panel or software. This solution of course, cannot differentiate so all the form is eliminated at scan time - displaying the form requires an electronic template to be first displayed on the screen.

Color scanners (normally 256 shades of red, green and blue) are now beginning to be used for forms processing. The advantages are

1. that the color shade of the background ink can be precisely-measured regardless of which color it is and dropped out
2. that the form data is easier to read and interpret regardless of how faint it is and the background color
3. image QC virtually gets eliminated
4. rescans can be virtually eliminated

The disadvantages are that:

1. there is a lot more data to deal with and deskew can be time consuming
2. there are no color based recognition engines available
3. frequently the resolution is not high enough to do recognition

Still despite these disadvantages, color scanning will soon become the normal method of operation.

Security Scrolls and Shadings.

Sometimes copies of forms may contain scrolls or shadings to prevent parts of a carbon copy from being read. This happens when a form is used for a dual purpose. For example, a survey company may utilize one multi-part form to get answers to questions for different customers. They block out irrelevant answers on second and subsequent copies by printing a security pattern over the irrelevant area. Scanning solutions can either remove these scroll marks - which designers need to be aware of, or the image can be clipped to remove the offending area. If it is left intact, it can cause compressed image sizes to increase dramatically.

Other forms deliberately contain security scrolls to prevent fraudulent copying of the information. Examples are stock certificates where the owner's share holdings are printed on top of security scrolls - to be able to read this information the security scrolls must be removed. Software can remove these scrolls with some difficulty.

Shading blocks. Sometimes forms are shaded in blocks to make them more easy to follow with the eye. An example is a large order entry form that contained many items on a sheet. To make it easy to align the eye, the designer had put inch wide blocks across the form similar to computer paper. This shading can cause interference with the image once it is converted to black and white.

Color images have the ability to remove the most complex security scrolls, even those printed in multiple colors.

Method of Printing the Form.

The baseline forms may be created by laser printer, ink-jet printer or printed by a professional printer. Whatever method is used they will vary somewhat. Even when they are printed by a professional printer and all of them look the same to the naked eye, they will still probably vary slightly. Within a print batch this is because of the normal movement of the paper through the press. Between re-printed batches printed at different times it may be affected by how the film is positioned, the press and the paper types.

A standard form such as the HCFA-1500 Medicare form is printed by a number of different printers to a defined specification. It all looks the same to the naked eye but the differences can be substantial.

Paper gets pulled and stretched and is affected by humidity. This can cause all the boxes to be slightly differently located in relationship to one another.

We are not talking about anything dramatic, but when a page is being scanned into 200 or 300 little dots to an inch and software is trying to precisely line up the dots to match a pre-scanned template, a line that is a fraction of an inch moved, or all the boxes shifted down, must be adjusted for by the software. The more sloppy the alignment, the more difficult the matching.

Colors can also vary from batch to batch. It is fairly normal for scanner manufacturers to publish the PMS colors that they drop out with the different light sources. If you do not have these, get them before you speak with the printer.

Design Methodology and Level of Control over Design.

Forms design can be either optimized to automatic recognition or optimized to be manually recognized. It is possible to design the form to carry both out successfully, but this requires some compromises on both sides. Forms designed by the user to be recognized and captured successfully have some key characteristics:

- Clear definition of what form it is.
- How it is registered how it may be deskewed or straightened in the computer.
- Clear and logical setting of where the information is located.
- Ease of extraction of the data.

All these issues can be addressed and optimized as a part of the overall design. However, if you do not have any control over the design of the form, because it is essentially someone else's form, then it may not be easy to recognize and process automatically.

Key Items that help:

Print an easy to recognize logo or pattern identifying the form. Some software that designs forms for recognition will create this mark that the software then recognizes.

- Print at least 2 and preferably 3 L shaped registration marks in the corners. These can be used to deskew the paper quickly and locate the start of it consistently within a bitmap.
- Separate check-off boxes clearly so that when a person checks off the box the mark does not potentially interfere with another adjoining mark.
- Print the actual form in pastel colored drop-out inks.
- Make boxes sufficiently large that instructional text does not get in the way.
- Put other instructions or legal text in rectagonal locations that can be cleanly identified and clipped from the overall image.
- Consider utilizing barcodes for a primary recognition. Choose a self checking barcode with a check digit. Make the thin bars

at least 1/50th of an inch wide (a rule of thumb is to leave 5 characters per inch space). Print the barcodes horizontally and leave a clean space around the bar.

- Print any text that should be captured in a clean clear font, preferably non-serif with a minimum size of 10 points
- Design the paper to make it 14lb or more with a straight sharp edge.

All these items will help improve the chances of automating the capture process.

Method of Completing the Form.

Forms are completed or filled in by many different people, in many different environments, utilizing many different processes.

At one extreme we have an individual equipped with a blunt pencil or a partially empty pen filling in little boxes of information in script handwriting. You may request the information to be clearly entered using a sharp pencil and block letters, but the individual may not comply. If you need the information more than he needs to complete the form - then you must process what you receive.

At the other end of the spectrum is the simple form printed with its data on a laser printer or ink-jet. The completed data is straight and is not interfered with by the boxes or instructions. In between there are many variations.

Control over Completion.

If you have control over the person completing the form, you have a better chance of them taking care when filling it in.

At this extreme, we have an individual completing a form requiring dense information as fast as he possibly can. An example is a complimentary trade magazine subscription completed on the trade show floor - some of these questionaires want 200 odd questions answered and all the individual wants to do is get his subscription and move on around the show. He has a little incentive to do a reasonable job, so the chances of success in this environment are extremely low. Another example where chance of success is low is where a form, which consists of mostly textual information, is completed by a member of the public whose literary skills are not well developed.

At the other extreme is the individual who is specifically rewarded for completing the form according to instructions or badly damaged if they don't. An example is an examination form - fill in the boxes neatly with a number 2 pencil - get the answers right and YOU WILL PASS — FILL IT IN BADLY or with the WRONG PENCIL and YOU WILL FAIL! What an incentive. These types of users tend to comply.

Most forms fall into a category between, but the more control over the individual completing the form that you have, the more chance of successful automation.

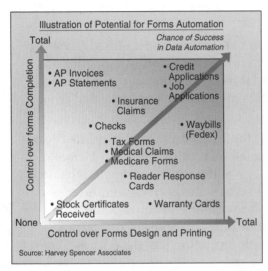

Figure 5.1 Illustration of Potential for Forms Automation

Content.

The content of the form affects automation success. For example an in-house survey of employees where the questions are fairly simple can be easily automated. If the form can be linked to one or more databases for validation or automatic pick-up of information, and if fields are known and can be automatically verified though a check digit and edit validations, the chances of automation success improve. Barcodes are a defined format that can be accurately converted and, in some cases, will self verify (see barcode recognition later). Check boxes can be an accurate way to gather data provided a validation is included (see check box recognition later).

Forms Processing vendors can help substantially with the design of a friendly paper form. In addition, there are now some software tools which have been developed to assist with designing friendly forms.

These graphically oriented tools help with field layout and spacing, with color selection, with constraining marks and with creating anchor points. Consider asking your vendor for one of these tools.

Electronic Forms

Defining

Electronic forms can be categorized into six main areas which are: impact printer replacement; print on demand; typewriter replacement; mail enabled; hybrid and Internet.

- Impact printer replacement electronic forms utilize a laser or other non-impact printer with forms printing capability to replace a line printer or dot matrix printer and pre-printed forms. Both form and data are printed simultaneously on plain paper.

- Print-On-Demand (POD) for forms replaces pre-printed forms which are used in very small quantities. Print-on-Demand systems maintain the image of the form (see creation below for information on how to create this) on the computer, and print it out as required. It saves long print runs and eliminates the need to throw out unused copies of pre-printed forms when some item of information like the address, layout, or telephone number changes.

- Typewriter replacement is a system used to make filling in forms easier. In some ways, the term typewriter replacement is a misnomer as few forms are now completed by typewriter. Perhaps a better term would be "constrained forms completion". A form is displayed on the screen of the PC and the operator is lead from one field (box) to another. The system allows boxes to get completed automatically through extensions, database look-ups and totals. Validation rules for alphabetic or numeric information, or specific edit pattern formats, such as dd-mm-yy, ensure the correct data is entered. An extension to this are products designed to accept a scanned image which is then used to create an electronic form. The lines are changed from pixel bitmaps to vector addresses and the text to true font equivalents. The form can then be edited or modified, completed and printed if necessary.

- Mail enabled forms are designed to be handled electronically. In fact they may never be printed. The forms remain as vector images with the data kept as computer data.

- Hybrid forms combine aspects from the other types of forms and may contain additional types of data such as a signature or stamp.

- Internet Forms are currently based on HTML, (HyperText Mark-Up Language) a textual based language which provides formatting information. The forms are displayed by a standard browser such as Netscape or Explorer and can be completed by the individual. The completed data is then sent to the requesting WEB server by clicking on a 'submit' button. This sends the data stream to server where it is processed through a CGI (Common Gateway Interface) script written in a language known as PERL and a response sent back to the sender. Internet based forms are being supplemented by a language called Java script that allows a certain amount of range type validations to occur before the data is transmitted. Java or Active-X control code is being downloaded to manage fields that require more validation or mathematical calculations.

The first three of these electronic forms types still provide a printed output which must be input by someone else. However, the creation method assists automatic forms processing since the form is straight, the text straight on the form, the text normally black, the paper white, and instructional text and boxes do not get in the way of informational text. Therefore, much image enhancement and pre-process can be eliminated.

Mail enabled forms can be distributed electronically and the data extracted from them without needing to pass through any recognition technologies if the recipient is using the same forms processing software. If not, then the form must be processed in the same way as a paper form i.e. recognized, a template applied to identify the fields, the data extracted as a bitmap pattern, and then recognition applied to it. Hybrid forms may contain both images and data and will need different types of processing to manage them. Internet forms can be supplemented with voice instructions, video or linkages to other pages for explanations. In the future, it will be possible to press a button on the form to get voice based instructions with an option to switch to a live 'help person'.

Creation

Electronic forms can be created through a forms drawing package normally included with a system, a standard drawing package, or through scanning an original form. Electronic forms are designed to let the user define a form that can be displayed, printed, sent by E-mail, or fax'd. Data can be entered manually or collected through a linkage to a standard database through linkages such as SQL or ODBC. Calculations and extensions can be automatic.

For example, a part number and quantity can be entered. The forms processing system will look up the part number in a database, extract and complete the detail, extract the price, calculate the extended price and total up the complete amount. The form is automatically stored or can be delivered through E-mail. A database is set up to allow any forms to be retrieved on the basis of keywords. While the form is in electronic format it can be manipulated and data extracted from it. Most systems have an export function to populate a database. But if interchange is required with people who do not have the electronic forms processing system installed that created the form and managed it, then they must receive it converted to a bitmap image either as paper or as a FAX.

Browser based HTML forms can reduce this problem, but if the recipient does not have the CGI script to understand the data, it is useless.

PDF & Others

PDF stands for Portable Document Format and was developed by Adobe Systems. It is a method of encoding a published document based on the PostScript language that was designed to solve the data interchange problem in electronic publishing. This occurs when the recipient of electronically published information does not possess the program which created it, preventing him from displaying it. A number of vendors have developed a standardized format capability which carries the formatting information with the document. Utilizing this type of format and a vendor supplied viewer (which can be normally acquired free from an on-line service), a form or other formatted information can be viewed and searched based on full text searching. As these formats become common and interfaces are built, electronic forms are becoming used more for business to business transactions.

The difference between this PDF format and a standard bitmapped image is that the data is mostly made up of computer understandable data. Therefore, the compression can be greater and mixed images (text and photographic) can be easily and efficiently encoded.

PDF has been expanded to display forms with validations based on Java Script. It has also been expanded to support version control and notations which can be useful for electronic forms as well as publishable documents.

XML

XML (Extensible Mark-Up Language) provides extensions to the Internet HTML to enable standardized formatting and understanding of specific fields to be carried with the forms data. The DTD or document type definition controls the overall format while XSL, which codifies style sheets allows the paragraph formatting to be carried separately from the document. XML will be used increasingly as a standardized format supported by most design and word processing systems. Therefore forms will be developed using XML and will be publishable either on paper or electronically. PDF and other formats will probably become XML compliant. MathML will complement this by providing mathematical instructions, while XML Schema will provide security control.

CHAPTER 6

Forms Delivery & Forms Capture

Forms that will be converted and filled out as paper, are delivered by one of four main methods:

- Directly
- By mail
- Electronically though the Internet or E-Mail
- FAX.

The method by which forms are delivered has an impact on whether they can be easily automated or not. The main issue is whether they can be easily recognized. Forms that are folded, stapled or on special types of paper will generally cause more problems than those on standard bond paper.

Readers of this book though should take any comments with some discretion. I recently came across a form which was printed on 24 lb paper, was very clean and had not been folded in the mail. It would not however, load through the scanner's autofeeder. We changed rollers and tried adjusting settings, but it still would not feed reliably. I suspected the loose weave of the paper as being the culprit, but we never were able to fully ascertain what caused the problem.

The conclusion is that there is no substitute for always trying out the equipment and software that is intended for production use with paper and images that are the closest to real life as possible in volumes and types.

Direct Delivery

These types of forms may come in internal mail or they may be handed over a counter or directly returned to an interviewer. Typically the form will be placed in a folder and may have an attachment stapled to it or a sticky note attached to it. The data preparation function must remove and potentially batch any attachments for scanning.

Mail Delivery

Most forms delivered in the mail need to be removed from envelopes and scanned. In some cases it may be necessary to scan the envelope too as the postmark may be important. Envelopes are a problem as windows can cause reflectance and the double thickness can cause jams or double feeds. Any envelopes that are sealed with a metal tag cannot be placed in an autofeeder. Mail delivered in envelopes is usually folded and the creases can cause shading in the image or problems with autofeeding.

Internet or E-Mail Delivery

E-Mail forms are the easiest to deal with, because the characteristics of a digital document make it easy to understand. However, an E-Mail form is probably mixed up with other types of mail within a batch of documents in an electronic "in-tray" and the form may have notations associated with it. It is probably simpler if you set up a separate in-box for forms or arrange for a data preparation person to tag the forms with a pre-defined code and sort them into separate areas.

FAX delivery

FAX'd forms are effectively remotely scanned forms. A Fax is a scanner which converts the paper line by line into a bitmap and compresses it in a format known as group/3 line by line.

The best results can be obtained from electronically created FAX's transmitted in FINE mode (which is 200x200 dots-per-inch), the boxes will be clearly delineated and the FAX will be straight - noise spots will be minimal. The worst results will result from paper FAX's sent in normal mode (app. 100 dpi vertical x 200 dpi horizontal) and received on a standard thermal FAX machine. All fax'd images must be converted into standard tiff group/4 compressed images before being able to process them with a forms management system.

The problems caused by faxes are as follows:

- the paper can get skewed in the sender's autofeeder leading to skewed text on straight paper at the recipient's site.

- the CCD on the sending FAX can be damaged leading to a continuous black line down one part of the form.
- noise from the FAX machine combined with telephone line noise can introduce a lot of spurious marks.
- the paper can get scrunched or sometimes elongated in the sending FAX's transport.
- 100 dpi vertical is not really adequate for good OCR or ICR recognition.
- some FAX paper is thermal which is thin and shiny. It cannot be easily scanned. Most scanner's autofeeders will not accept this type of paper and it curls making it difficult to place on a flatbed. To scan it you usually need to photocopy it first. This introduces another process that can cause problems.

FAX's received electronically can usually be routed over the network, converted into group/4 and imported directly into an imaging capture system as batches. The system then treats them as any other scanned document.

Manual Forms Handling

Paper forms are completed and managed in many different environments. For example in a warehouse, proof of delivery documents may get torn or otherwise damaged. They may even get dropped and marked as someone treads on them with a dirty boot!

Overall the more manual processes that a form is subjected to in an uncontrolled environment - particularly oudoors (as with Bills of Lading and Shipping Documents) or in a production shop - the more subject to damage a form becomes. If you can find a way to reduce handling or at minimum print on thicker paper, the chances of success will rise.

Timeliness

Forms may contain information which must be processed immediately, or they may not be very time sensitive at all. An example of a time sensitive form is one that must be processed in order for something to happen or where money is tied to the immediacy of the document - a hospital admissions form is one example, another example would be credit card receipts. Shipping documents are reasonably time sensitive as customer service needs access to the forms as soon as the shipper is returned. Warranty cards however, may not be as time sensitive since it will not make much difference if the information is captured the same day it is received or a couple of days later.

Timeliness of information has an effect on price. In the case of a service bureau, data that can be fitted around other priorities can usually be priced more competitively than that which is required with a fast turn around.

CHAPTER 7

Before Scanning Paper Forms

Preparation and Batching

Spending time on document preparation is probably one of the most rewarding activities in improving overall throughput. As already stated, forms are delivered

- in multiple different formats
- in multiple sizes
- in multiple different qualities
- on many different types of paper
- folded in envelopes
- with multiple pages stapled together

To be scanned, the envelopes must be opened, pages must be removed from the envelope, decisions made regarding what gets scanned (e.g. envelopes, single or two sides), staples removed and any repairs to the paper made. Depending on the application, the forms can be separated into process routing batches and controllable segments, which makes scanning and recognition simpler and therefore more automatic.

Although not always possible, scanning performance can be substantially improved by sorting the forms by size, by paper thickness and by image quality. Batches can then be set up to deal with these. If some images are going to need specialized treatment with algorithms to remove backgrounds (see OCR pre-process in a later chapter), then it is much faster to outsort them before scanning.

The first activity is to prepare the documents. This involves getting them out from envelopes, smoothing out the folds, removing staples and repairing any tears with tape. At the same time batch control sheets can be added.

Small Sheets

Some scanners and most autofeeders are not tolerant of mixed batches of small and large paper as they will tend to skew the small sheets - this is particularly prevalent with center feeders. The guides must be set for the widest paper and then the smaller sheets tend to skew badly. Some forms applications — for instance claims processing require the user to maintain the integrity of the original paperwork. Others — for instance single sheet forms such as tickets or invoices — can be processed in any order. Sometimes controls are needed to put the batches back together or sort the images electronically after scanning.

There are five possible ways to deal with small papers depending on the application:

1. Paste smaller documents onto standard letter sized sheets. It is preferable to mark paste-up sheets with faint dotted lines in different sized boxes so that the paper can be lined up straight while pasting. Note: roller transports on scanners and autofeeders do not like pasted documents. The edges can get folded back and in extreme cases they can jam the feeder.

2. Batch the documents into sizes so that the guide bars can be set to the right size. This is not always possible when dealing with homogeneous groups of documents such as expense statements attached to a pile of receipts.

3. Utilize a "carrier". Some scanner manufacturers provide a transparent envelope known as a carrier. The smaller or very thin document can be placed into this envelope which effectively makes it a standard letter size.

4. Utilize the flatbed. In this case, the lid of a flatbed scanner must be lifted and the paper placed flat on the glass before scanning. If the operators are trying to achieve production, they may be so quick placing the paper that it ends up very skewed. As very skewed paper will reduce OCR and other recognition accuracy, you may want to tie an operator's compensation partially to OCR accuracy levels or numbers of paper unskewed.

5. Utilize a scanner that checks for a straight piece of paper. Some scanners feature a sensor which checks for a straight piece of paper when manually feeding. If it isn't straight, then it is returned to the operator for re-feeding.

There is some compatibility now available to deskew electronically in grayscale. By using the tonal information, grayscale deskew can almost eliminate the need to worry about skew. (see also Chapter 10 for discussion on electronic deskew and shading removal).

Varying Thickness

Scanner's autofeeders will often cause double feeds when asked to feed different thicknesses of papers. When a transition from thick to thin occurs, the rollers will open to accommodate the thick paper which will allow the thin sheet to slip in as well. One way to avoid this is to batch documents into separate thick and thin batches. Some scanners have dealt reasonably with this problem by increasing the interpage gap. This creates time for the rollers to close down but has a performance impact. Most higher performance scanners now include double-feed detection (see chapter on scanners for more on this).

Patches and Coded Labels

While preparing pages for scanning, automatic routing or indexing, labels can be attached to the pages by the preparer. These labels are often placed on the page through the use of patches or stuck on bar-coded labels. Barcodes are reliable and easy to read automatically. In fact, some scanners provide a barcode reader inside them which can read the bars as the paper passes underneath while others include bar-code recognition software within their capture solution. Bar-coded patches can be created to enable almost any alphabetic or numeric sets of characters, and stuck on in any unused area provided there is enough space on the form (see barcode recognition later).

Batching

Forms based documents can be batched into two separate categories:

- Single page forms.
- Multiple page forms (including two-sided forms)

Batching Single Page Forms:

Batch control requires the entry of batch control numbers before scanning the batch and a variety of vendors are using OCR and bar-code recognition to automate this. Some of the companies whose background was in using microfilmers for Service Bureau operation developed the concept of predefined specialized Patch Cards that were recognized automatically with ultraviolet light to control batches. While the ultra-violet light approach is a proprietary feature, this technique has been adopted with specially coded pages that can be read optically (see fig. 7.1).

A single page form application such as single-sided warranty cards can be batched into a controllable group - say groups of 100 pages. A recognizable batch control sheet is then added in front of each batch. This normally contains a large barcode or other marking that can be easily recognised by the software or hardware within the scanner. This requires specialized software to recognize, so it is usually defined by the software supplier.

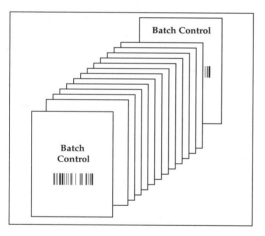

Figure 7.1
Typical batch of papers

The scanner scans the batch control sheet and then starts counting the pages following. If the count hits 99 or less when the next batch control sheet is reached at the start of another batch, the operator knows there has been one or more double-feeds. Then he must identify which sheets were double fed by matching the images on a screen to the physical paper. This repair time can be substantial, so scanner manufacturers have incorporated endorsers or imprinters within their scanners. These devices print an identification, usually under programmatic control, on the paper as it is scanned. Typically the software will print a date, time and sequence number on each sheet. The advantages are that:

1. if there is a double feed, the operator can easily identify where it was as the second sheet will not be printed on

2. The paper can be easily resorted into order for filing if required

3. The image contains the endorsement code if the endorsement is before scanning, so if the physical paper is required, it can be easily located from the image.

Autofeeders usually take a limited number of sheets - normally about one-and-a-half inches high on bottom feeders (more than that causes the weight of the paper to prevent feeding). As the pages are scanned, more pages can be added to the top of the stack.

Batching Multiple Page Forms:

Multi-level Batches.

The microfilm industry developed a system of 'blips' or small black squares that were located at the top of the film. These were defined in three different sizes which represented three different levels of index - these could be roughly defined as

- topic
- folder
- document

When a document or page was sought, the retrieval database software would understand which blip number was associated with a look-up. It would then quickly scan through the roll of film counting the blips and looking for the appropriate one. This principle has been adopted in some document processing applications utilizing different batch control sheets to define the levels.

A multi-page form application, particularly if the number of pages is not pre-defined, may utilize different levels of batch control.

- One to identify groups of documents

- A lower level as a separator for each form.

In this instance the software will typically create subdirectories on the disk based on the levels.

An example in the insurance business might work as follows. A typical insurance claim consists of a claim followed by supporting documentation. The first batch control sheet will be identified as a master batch control through specific barcode or OCR'able field. This will create a directory labeled with the 'claim number' which will be retrieved from the barcode or OCR field on the batch control sheet. The pages that follow will then be placed in that directory as separate image files forming an electronic 'folder' for that claim. In some cases, the user may add lower level control sheets to identify a 'type of document' (e.g. correspondence, doctors reports, etc.).

Two-Sided Forms

A scanner treats two-sided forms as if they were two separate pages and normally captures two separate images. If some forms have data on the back, which needs to be captured, and others do not. Then it will improve productivity and reduce storage, if the forms with and without back-sides are separated at preparation time.

Data Preparation Productivity

Studies have shown that a typical clerical worker can prepare approximately 750-1,000 sheets of paper per hour for scanning depending on the amount of repair required.

One thousand pages/hr includes removal from envelopes, separating the sheets, counting and batching clean and undamaged papers.

The seven hundred and fifty figure includes some level of repair of torn pages - using scotch tape for example, and pasting up small documents onto larger letter sized sheets.

Adding patch stickers with barcodes would add time, which will vary depending on

- whether there is a fixed location that it can be placed on,
- whether the data preparer has to look for a space that it can be placed on and
- whether the codes are generic or specific

Note that any high levels of repair can be very time consuming, so plan to organize the form so as to minimize it.

Above all do not think preparation is a waste of time.

The time spent on data preparation will pay back many times over in less jams or double feeds in the scanner, less operator data entry, and improved performance.

Forms Management & WorkFlow

Workflow is a term which means the management of office transactions. It has been used to mean anything from sending a form from one person to another via E-Mail, to the complete automatic distribution and control of a form using specific rules. Although workflow can be performed within one computer in small applications, it is invariably linked with networks. It can be invoked as a part of the capture process, or it can be used subsequently when processing a form. Depending on the application, the way that forms are processed consists of:

- pre-defined workflow,
- knowledge based workflow,
- ad-hoc workflow,
- a mixture of all of them.

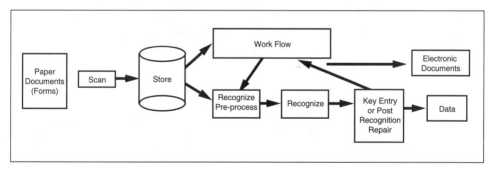

Figure 8.1
Typical forms processing workflow

For this discussion we can separate workflow into two main different categories:

- Input Workflow and
- Management Workflow.

Input Workflow

Input Workflow is a process control workflow designed specifically to handle the movement of, and the processing involved, in capturing large numbers of forms or documents. In the case of a forms processing system, it is designed to optimize the scanning and reading of the data from those forms. In the case of document capture, it is designed to optimize the scanning and indexing of documents.

The basic elements, which are similar in both cases are:

- scanning
- pre-recognition processing
- recognition
- validation of the data
- repair
- export

Each element requires different amounts of manual operator time and attention, and different levels of processor power.

Scanning a page on a high speed document scanner, typically takes less than one to about three seconds, to transport and convert one or both sides of an average letter sized piece of paper, to a black and white digital image, at 200 or 300 dots-per-inch resolution. Grayscale takes a similar time for the digitization process, but outputting the information, which tends to be a lot larger, involves more time. Color scanning usually either takes more time or involves reduced resolution.

Enhancing the image and preparing it or recognition can take from a fraction of a second to several seconds, depending on the amount of work that must be carried out. It also can be affected by whether the enhancement is hardware assisted or not, as these processes tend to be processor intensive. If there is no hardware assist, then faster processors will improve the speed of this process.

Recognition, which may involve multiple different recognition engines, can likewise take a fraction of a second to many seconds,

depending on the amount, location and complexity of the data and processor used.

Validation is affected by the size of data tables to be looked up.

Both repair and key verification are manual keyboard processes that are labor intensive. A typical specialized key entry operator will key mixed alpha-numeric data at around 10,000 characters/hour. But this can vary from as little as 7,000 characters/hour to as much as 16,000 characters/hour depending on:

- how much data is simple numeric data vs how much is alphanumeric.
- where the the data entry fields are located .
- how much data is handwriting.
- how much visual interpretation the operator must perform before keying.

Once the forms are scanned, the image files will generally be located on a network server in separate directories according to batches. Input workflow manages the flow of these batches of documents through the input capture process, allocating and spreading them across systems as efficiently as possible. As much of the recognition pre-process and recognition functions can be managed unattended, some workflow will re-schedule and reconfigure systems to handle more than one function.

Some types of applications or forms need to be processed and interpreted by clerical staff (e.g loan applications, tax returns, or insurance claims). Other types of forms (e.g. surveys or warranties) merely need the information to be captured from them and after that stage the form has finished its useful life apart from being needed for proof in some cases (for example: subscriptions to trade journals need proof of signature for audit purposes).

If a form needs to be looked at and managed by clerical staff, it may be routed via management workflow to control its flow, its dependencies and the connection to and flow of all associated information through an organization.

Pre-defined workflow provides for the defined rules to drive the electronic documents through the forms processing system and make decisions regarding routing them. Pre-defined workflow can be very simple.

For example batches can be set up as routing elements or a person must "tag" the images with an identification or routing code. This can be manual using the keyboard or via an automatically readable code such as a barcode stuck on the paper.

It can also be very complex, in which case it tends to get combined with knowledge based workflow.

Knowledge based work-flow is used to route documents based on the content of the document. In this case forms must be captured and interpreted in order to trigger the routing.

In the capture process knowledge of the potential accuracy of a scan on the input process is able to sort out potentially badly scanned documents and route those for re-scanning or to an electronic repair process.

A document processing example is in mortgage loan processing. Loan documents are received with attached supporting documentation. Each loan request type is set up with rules to control the routing depending on parameters laid down by the lending organization. The application must be looked at to identify its parameters. It must then be checked for completeness and any missing data requested and then be routed to the appropriate people or through the appropriate processes. It may be routed to a supervisor if over a certain value or if the loan has some unusual aspects to it, or it may flow through a main process without "exceptions". After processing the form and associated documentation is filed (i.e. stored on archival media)..

Workflow is a demand "push". A clerical member of staff processes or looks at a form and determines that someone else must see it. This could happen because the form fits the knowledge based rules, but does not pass a manual check. Typically E-Mail will be used to carry out this function.

In a forms entry environment supervisor overrides to the process are essential. For example, a supervisor monitoring the process speeds may note key entry speeds in relationship to a specific operator and specific form type. The supervisor may then route batches of similar forms to that operator's queues, possibly overriding the process control. Alternatively a new job may arrive which requires priority processing so the supervisor may need to dynamically change the job sequences.

Some workflow allows users to apply a financial factor to each process so that costs associated with processing each form type can be measured. It's one way that workflow can assist in identifying and improving process efficiency in the office.

Getting the paper into the system

Scanning

Forms processing is affected dramatically by the scanner selected. The transport (i.e. the method of moving the paper) is the item that affects cost and capability the most. All general purpose scanners that handle mixed sizes of forms manage the paper flat like a photocopier does. Higher speed scanners, which manage the paper on edge, are designed for smaller documents which have a certain homogeneity in weight and size. These are used in applications such as remittance advices and checks. The paper can be moved faster and gravity controls the level of paper skew.

General purpose forms scanners, which can handle both small paper and letter sized sheets, vary dramatically from simple roller fed transports to complex devices that utilize a vacuum to hold the paper down against a transport. The cheapest scanners are roller fed, using foam based rollers to sandwich a piece of paper and move it past the light source and CCD (see fig 9.1). Generally a roller transport can only move a letter sized page at less than 10 inches/second, which works out to a scan time of greater than one and a half seconds for a letter-sized sheet.

The medium speed transports utilize belts or drums and can move paper at least twice the speed of rollers. The high speed scanners use vacuums to hold the paper flat while using belts and rollers to move it. These types of scanners can digitize a piece of paper in less than half a second.

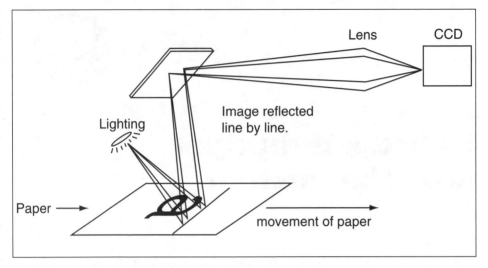

Figure 9.1 How a scanner works

But the transport is only part of the issue. All forms processing soft-ware relies on clean images that can be easily recognized. Scanners optimized for document imaging and forms processing contain hard-ware designed to sharpen and improve image quality, but the software must control this hardware.

How to Select a Scanner

Selecting a scanner suitable for forms processing can be confusing. There are over 100 different scanners on the market each of which can perform the basic job of converting a paper page to a bitmap repre-sentation.

Desktop Publishing vs. Document Imaging

A major area of confusion for the newcomer to document imaging systems is in differentiating between desktop publishing scanners and those designed for document imaging. The main differences between desktop publishing and document imaging are in:

- Speed of scanning (i.e. the time it takes to convert a letter sized piece of paper into a digital representation).

- Resolution (i.e. the number of dots-per-inch - known as dpi, that the image of the page is broken up into).

Forms processing is a technology for routing, recognizing, filing and retrieval of forms. As discussed earlier, most of these papers are bi-tonal (i.e. they contain a background and a foreground in a different color). Most of the required content is textual in nature, printed in fairly large type, and any backgrounds represent interference with the information that is really required.

The functions needed for document imaging and forms processing scanners are:

- High speed scanning - typically less than one, to three seconds per page

  Forms processing normally requires scanning several hundred to several thousand forms per day and the forms need to get converted as fast as possible, so that they can be processed. Anything less than a photocopier's speed is unacceptable.

- A maximum resolution of 300 dpi with 200 dpi capability.

  Text over 8 point size can be resolved for viewing and printing at 200 dpi although it may be difficult to OCR it. It is rare that users have a requirement to resolve less than 8 point text and 8 point text tends to be boilerplate or instructional text and therefore not required for OCR purposes. Note however, that a resolution of 100 dpi is adequate for manual key entry if the image is scanned in grayscale or color

- Black and White scanning (known as bitonal).

  In principle forms do not need to be captured in color. The color information does not mean anything and it uses a lot of storage and network bandwidth. Users are used to black and white images from black and white photocopies and faxes. Also recognition technologies use black and white image patterns. However, color scanning can convey some advantages. It allows for easy drop-out of forms. It provides for less need for image quality control and rescans. It can provide for easy keying of certain images such as the red on blue airline tickets. AND it provides a true rendering of the original image. Some scanners provide for both color and black and white output from the same form for this reason.

- Image Sharpening.

  Forms contain lines and text. To accurately process them, a scanner needs to sharpen edges of characters and remove shadings and background noise. This is exactly the opposite functionality from that required by a scanner of photographs.

Compound Forms with Photographs and Text

Some forms applications (for example passport applications) have a photograph as well as textual information. This causes a problem on black and white scanners as the photograph gets so badly contrasted that it is frequently unrecognizable. Some scanners automatically differentiate photographs and dither or screen it which provides a reasonable rendering. However, the subtle shadings provided by grayscale or color are really needed for these types of applications. (see below under image reproduction for more on this)

Figure 9.2
High Speed Document Scanner

Source Kodak

Scanning Speed

Most businesses deal with hundreds or even thousands of forms a day. Desktop publishing requires the scanning of a very few items in a day, but they must be high quality. The process often involves placing the original artwork on a flatbed piece of glass, invoking a scan and viewing the resulting image. If it is badly contrasted or too light or dark, the operator will make an adjustment and re-scan the item. It may take as much as 10 or 20 minutes to get a GOOD image from a poor piece of artwork, but because the operator is typically only scan-ning 5 — 10 — or 20 pieces in a day, the time is not as important as the quality. The business forms processing system scanning 1,000 or 20,000 pages a day wants to put the paper in an auto-feeder and walk away, finishing scanning as fast as possible without having to stop to make adjustments.

Desktop publishing scanners typically take 5-20 seconds to scan a page, whereas document imaging scanners typically take 2-3 seconds or less per page.

Scanners quoted speed parameters can be misleading though. Scanner manufacturers usually quote scanning speeds based on 200 dpi for a letter sized page scanned in landscape mode so as to mini-mize the paper length. But often you cannot scan in landscape or it takes time to rotate the image 90 degrees. Sometimes paper is larger than letter size and often smaller. Some software requires 300 dpi to get good resolution. So to analyze your actual scanning times you need to look at the transport speed in inches/second at a specific res-olution, find out the gap between pages and calculate times based on the anticipated mix of documents. For example if you are scanning 4 inch long sheets at 11 inches/second at 200 dpi and the interpage gap is 1.5 inches then each page will take a half second. Scanning at 300 dpi will reduce the speed by 2/3rds.

Resolution

Type sizes on business documents are created large enough to read easily. Typically this is typewriter sizes which are 10 point (elite) or 12 point (standard). Occasionally disclaimers or instructional text will be printed in 8 point size and sometimes headlines or titles will be in much larger fonts. However, all these types of information can be eas-ily captured at 200 dots-per-inch (dpi).

The higher resolution 300 dpi which will take longer to scan, increase image sizes, and degrade network performance, is not usually required for forms processing although it can make improvements with OCR in some cases. Interestingly some recognition works less well at higher resolutions as OCR is usually optimized for 300 dpi although it can work adequately at 200.

Desktop publishing scanners often go up to 600 or 1200 dpi which is definitely not needed, it slows down the scan speed and increases file sizes. This leads to worse network performance and much higher disk storage requirements.

Use of GrayScale

Bitonal means black and white. Grayscale is the term used to convey the carrying of shades of information. When you have more shades, the eye requires less resolution to read the document. Normally scanners have the ability to output 8 bits/pixel of gray which equates to 256 shades. This gray information can be useful to create intelligent thresholding. Thresholding is the conversion from shades of information into black or white (see image enhancement later in this chapter).

Use of Color

Although many forms feature color, it is usually used to draw lines, display corporate artwork such as logos or in the color of the paper to differentiate the recipient. From a document imaging and management standpoint, color and shades are usually irrelevant. Most document imaging scanners are bi-tonal. They will convert the page to either black or white dots. However, some are available that scan in color and convert to bi-tonal based on the color information. The attractiveness of this is that different parts of a colored form can be set to be captured and others dropped out based on their shades.

Desktop publishing scanners utilize many bits of information to capture shades of color - typically they will separate the image into Red, Green or Blue and utilize at least 8 bits of information to portray 256 shades of each. Therefore each dot has 24 bits (3 bytes) of information associated with it. Even with compression this represents very big and, for forms processing, unnecessary file sizes.

Most forms processing and document imaging systems compress images using a variation on FAX compression known as Group/4. This is a bitonal compression that will reduce the file size of an an average form to one tenth the original size and guarantees no loss of information.

Color images tend to be compressed into using an algorithm developed for photographs called JPEG. This format compresses based on coding subtle changes to the colors and therefore potentially loses some information. If the user allows for an approximate 10% loss of color shades, he will probably achieve a 90% reduction in color or grayscale image sizes and the form will be perfectly readable by humans.

As color is increasingly used in document storage and retrieval systems forms processing systems will be adapted to use color images.

Image Reproduction

Reproduction of photographic information has different requirements than text or artwork. Essentially the more shades of color or gray that are captured from a photograph the better the photograph will appear visually. Shades are more important than resolution. With text or line-art, the opposite is true. The characters or lines need to be sharply delineated - shades will make them look fuzzy and adversely affect OCR or other recognition accuracy. Document imaging scanners are optimized to sharpen lines adding black dots to complete lines, or deleting them where they are extraneous. They can also track the backgrounds and varying contrasts, automatically adjusting to create the sharp contrast that is needed for good recognition to occur.

Scanners work on the basis of moving a page line by line past a charge-coupled-device (known as CCD array) which emits an electrical signal which varies in strength depending on the amount of light shone on it (see figure 9.1). To create this signal, a bright light, normally provided by high intensity fluorescent bulbs, is shone on the paper and reflected onto the CCD array. White parts of the document cause high reflectance, black, red, or dark colored causes none.

All scanners work similarly but the pricing of document imaging scanners varies from less than $1,000 to over $100,000. It's a big difference, so some discussion on the differences that occur seems appropriate.

Scanners with dynamic or adaptive thresholding and convolution filters enhance and improve the images. Dynamic thresholding measures the overall background of the image and adjusts for the contrast. Adaptive thresholding, which is standard on most document imaging scanners, adjusts for varying backgrounds. Convolution filters sharpen the edges of the characters (see figure 9.4).

Although much of this process can be handled downstream as a part of recognition pre-process, scanners can work on a much wider amount of information. As a page is scanned, the CCD puts out an analog signal which is digitized into a grayscale rendering. This grayscale is then manipulated to sharpen the image before converting it to black or white (see figure 9.3). Once that black or white decision is made, the supporting grayscale information is usually permanently lost (although some scanners have the ability to output both black and white and grayscale concurrently). Downstream recognition pre-process works on the basis of shapes of characters or lines (for more information see "recognition pre-process").

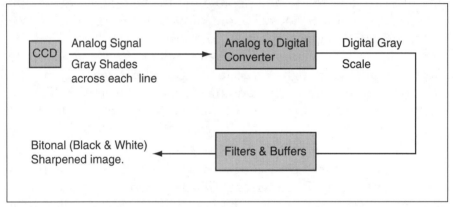

Figure 9.3
Internal Scanner Processing

Autofeeders work better with paper weights that are over 14lb (i.e. not the very thin stuff). FAX's work better if they are sent in fine mode which is 200 dots-per-inch (dpi) resolution as opposed to normal which is app. 200x100 dpi. While some support a new version of TWAIN most vendors support ISIS, a standardized interface which allows interface to most of the major document imaging scanners used today. Batch controls usually consist of a control page located in-front of either a defined number of pages (normally used when processing single paged forms), or may identify a logical grouping of forms and associated information (e.g. an insurance claim). The batch control sheet may contain validation information such as batch totals to be OCR'd and/or a barcoded reference number.

Scanning in batches of papers is helped if controls ensure that pages do not double feed and have an audit trail number printed on them using the scanner's endorser/imprinter. Scanners will always double feed if fast transitions occur from thick to thin paper as will happen in mixed forms - high end systems always seek to protect from this.

Two-Sided Forms

Some forms are two-sided. Two-sided scanners scan both sides in one pass of the paper, creating two separate images. These scanners cannot automatically differentiate between a blank side and a completed side. Although some vendors have incorporated some simple algorithms to assess the amount of information captured - if it is below a certain threshold then the page is assumed to be blank. Depending on the application this may or may not be useful as it could assume that a mostly blank page containing a signature only was blank and throw it away. We suggest that any suspect pages be manually reviewed before deletion.

In deciding whether to purchase a two-sided scanner, a good rule of thumb is that if more than 10% of your pages are two-sided you can cost justify a two-sided (or duplex) scanner. An increasing number of document imaging scanners are duplex.

Some manufacturers of scanning software provide a capability to scan the first side as a batch of papers and then turn them over and scan the other side. They then electronically paste the two sides together. If you are considering scanning two-sided pages do not think that a single sided scanner with software to paste the back and front sides together will do. All scanners double feed or jam at some time or another. Unless you have very few double-sided forms, the repair time involved (i.e. the correction for a double feed, jam or skew on one side which does not occur on the other) does not make it worthwhile.

Recognizing Double Feeds and Correction

As mentioned above, all scanner autofeeders will double feed at some time or another. Double feeds cannot always be recognized.

Some scanners feature double-feed detection which either stops the scanner or informs the operator after the batch is scanned. Double-feed detection has been implemented in four different ways:

1. an infra-red light is shone through the documents and if a thicker source is detected it stops the scanner.

2. a vacuum tries to pull two pages apart, triggering a micro-switch if it succeeds

3. an ultra-sonic signal is bounced off the paper and if the echo changes it detects and double-feed.

4. software looks at the shape of the paper and if it finds it is not a rectangle, it highlights the page and in some cases can stop the scanner.

All these methods have advantages and disadvantages, but if the scanner does not feature double feed detection, the operator will probably not see it occur.

If the scanner does not stop, or a batch contains a fixed number of pages, then the system can identify that a double feed has occurred, but not where. The operator then must go through a time consuming task of matching pages to images to identify the missed page. One way of reducing this effort, while providing a long-term audit trail of when specific documents were scanned, is to use an imprinter.

Imprinters which are now offered with most high speed scanners, use an ink jet to spray characters on the front of the page under program control immediately before or after it is scanned. Typically these are used to print a date, time and sequential number on the paper as it is scanned. Pre-scan imprinters can be manually moved to ensure that they do not overprint critical information needed for recognition. Imprinters help a double fed sheet to be easily found, as the page that does not have a number printed on it will not have been scanned. A secondary benefit of pre-scan imprinters is that the image of the page will contain the imprinted characters. If the physical paper is not destroyed but filed in some off-site low cost location in date and sequential number order, it can be easily retrieved from the image using the imprinted filing code.

Beware of so called "electronic endorsers" which place a bitmap image of an endorsement on the image, but nothing on the paper. This type of operation will not provide you with a physical audit trail. Therefore you cannot find un-scanned papers and it cannot provide an audit trail back to the actual paper record should you need it.

A similar device borrowed from check proof encoders, which is not as utilitarian, is an endorser. This literally stamps a fixed endorsement onto the back of the page as it is scanned. It provides the proof of scanning, but does not provide the same level of audit trail.

Image Quality Control

Quality Control problems with the image occur from differing sources:

- A page is inadvertently placed upside down.
- A landscaped page is placed in the stack.
- A page is accidentally reversed in the stack.
- The image on the form is faint and needs contrast adjustment.
- The image is too dark and needs to be lightened.
- The background or a mark interferes with the image.

Originally document scanning systems allowed the operator to control the scanning. He or she would release each page from the stack into the autofeeder, view it and accept it. The problem is that this ties up the operator - one per scanner - reducing the effective scanning time to around 5 seconds per page. To effectively scan batches of pages, the scanner should be partially unattended, if documents are homogeneous, and the scanner can be run at full speed.

Forms can be displayed, as they are scanned and viewed, for correct direction and quality. However if running in a high volume batch environment, it is difficult to react to a bad scan in time when a new image is being displayed every 2-3 seconds or less. Therefore most systems handle image quality control by displaying every nth (normally 10th) image on the screen or by displaying thumbprints. If there is an overall image problem, the operator can slow up the batch process and start viewing and releasing each image. In the meantime, the potentially problematic batch can be routed to a repair queue for a separate operator to scan each page, view it and, if necessary, apply different image enhancement algorithms to it. Alternatively the operator can stop the process and repair the problem image.

Images that are rotated can be accepted and dealt with later as a separate repair process although image compression will be damaged if the image is landscaped. Pages that have been reversed are a problem if the scanner is a single-sided scanner. The only real way to deal with this is to tighten up the preparation process, but software can identify a potentially blank image and route it to a "questionable" queue for viewing and possible action.

There are now some automated quality control and improvement tools appearing on the marketplace. These include:

- the ability to automatically identify a landscaped document and rotate it 90 degrees.
- the ability to automatically identify an upside down document and rotate it 180 degrees.
- the ability to identify if an image is potentially badly scanned so that it may be flagged for subsequent re-scan.

After scanning, the documents are collected into subdirectories, for pre-recognition.

Image Enhancement at the scanner.

Poor quality originals can be improved while scanning. Scanners designed for document imaging, typically include or offer optional electronics to improve and sharpen the image. Image enhancement falls into two main categories:

- Background removal
- Image sharpening.

Background removal is generally known as thresholding which automatically adjusts the contrast and the differential between darker - presumably foreground, and lighter - presumably background. Adaptive thresholding, which is the most sophisticated method, adjusts on a continuous basis across the whole image. This is particularly useful for forms processing as the colors or shadings often found will otherwise interfere with the image.

Image sharpening takes the foreground information and looks at each dot or pixel in relationship to the adjoining one(s). Then based on the theory that a valid character, as opposed to a speckle noise character, always has adjoining dots (don't forget that at 200 dpi scan we are dealing with 1/200th of an inch of the original paper per dot), the system will add back dots (see fig. 9.4).

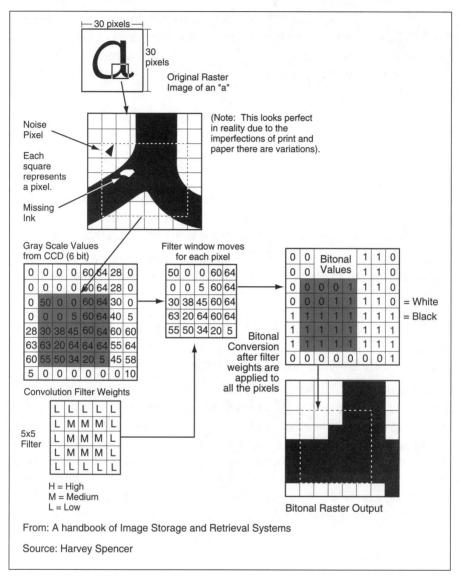

Figure 9.4 Simple example of image sharpening 5x5 pixel convolution filter.

These systems use different weighting algorithms and different numbers of adjoining pixels to make this decision. With some scanners the algorithm is user selectable for a batch of documents. This is a good idea if you know the outcome, but the problem is that image enhancement is a specific to a type of image object that the user wants (i.e. optimized to OCR, or optimized to line drawings). Forms are particularly difficult as they often contain different types of objects (i.e. a form may contain printed text, hand writing, barcodes and check marks) so some of these enhancements can degrade parts of the image in certain circumstances.

Tip: Before committing to a certain type of image enhancement check out the image quality in conjunction with your recognition engine(s).

Batch Scanning

Once the batches of pages are sorted into multi and single sheet pages with batch controls separating the sheets, then they can be placed in the autofeeder or fed sheet by sheet into a manual feed.

Manual feed is the most effective way to feed sheets that vary dramatically in quality and thickness, and rather than scanning the extra batch control sheet, it may be preferable to use the batch sheet to indicate to the operator that he or she must press a key to denote a new batch. Most manual or semi-manual feeds are designed to allow the operator to be seated in front of the scanner and the paper is returned to the front for easy unloading.

Autofeeders either feed the paper facing down from the bottom of the pile or feed face-up. Face-down is preferable for documents that are of similar weights as the pile can be added to while scanning, but the number of sheets is normally restricted by the weight of the pile to around 50-100 sheets maximum - i.e. about 2-3 minutes of scanning. The stacker also is normally limited and must be emptied continuously, although some users remove the output tray and substitute a deep catch box on the floor that the paper can fall into.

Face-up feeders will normally accept a much larger stack of paper than face down ones.

Quality Control

Images can be checked individually for image quality, but this is time consuming. There are two ways to operate in this way -

1. View each image and accept it if OK. If not, remove it from the out tray and re-scan it with different settings. This is very inefficient. By the time the image is displayed and an operator

has viewed the image and released it 10-15 seconds can have elapsed.

2. View each image at smaller resolution but keep the scanner scanning. The problem with this is that by the time the operator has realized that there is a poor image, the next two pages have been scanned. All three pages must be pulled from the output stack and re-scanned or better the bad page can be removed from the stack and placed in separate QC pile for re-scanning later - possibly with different algorithms applied to enhance the image or remove a difficult background.

In a batch scanning mode, the operator will load a stack of paper and possibly be able to manage two scanners. Although this is theoretically possible, I have found that in mixed document environments, it doesn't work well.

Batch scanning quality control can operate in two ways:

1. Every tenth page is displayed and it is assumed that the intervening pages are OK. This works well if all the pages and images are pretty much the same in quality and type, and the software automatically recognizes upside down documents and flips them. This method of operation works best if you batch into different paper and quality types.

2. Scanning takes place pretty much unattended. No images are displayed while scanning. The images and paper are then passed to one or more QC operators who flip through the pages checking for good image quality during which time they can attach a primary index or routing code. Sometimes multiple images are displayed so that the operator can quickly scan through them looking for quality problems. This method works best if there are several levels of automated quality control in the system: identification of upside down, landscaped and blank sheet suspects (occurs with a two-sided scanner on single sided sheets).

3. Scanner Maintenance
   In high volume situations, except to blow dust out from the scanner regularly. Expect to change rollers. And expect to change lamps every now and then (see manufacturer's specifications).

CHAPTER 10

Preparing for Automated Recognition - Pre-process

Reversed Images

Some types of forms such as birth certificates are printed in reverse i.e they are a print of a negative with the background black and the text white. Although these types of forms can be recognized, compressing them and storing them is inefficient and so it is preferable to convert them as soon as possible. This is a simple and fast task as black pixels are stored as binary 1's and white pixels as binary 0's. If the pre-batching separated these types of forms then this activity can be carried out either by the scanner or as a batch process. I prefer using a batch process, unless there is a batch control sheet which automatically changes the scanner's mode, as this is one less task for the operator to perform.

Rotation

Forms can be placed in a stack upside down or may be landscaped. Some two-sided forms are printed in a "tumble" format, which means that instead of turning the page over right to left, it must be turned over top toi bottom. In this case, the second side will always be upside down. In all these cases the images must be rotated 90 or 180 degrees. 180 degree rotation is simple and quick, but 90 degree rotation is often time consuming and some data can get lost. Hardware assisted image rotation can rotate images 90 degrees in less than one second. This function is often carried out while scanning.

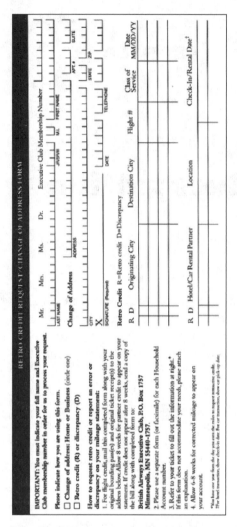

Figure 10.1
Example of landscape scanning

Registration

There are three problems that occur with a scanned image:

- The scanners's rollers pick up at slightly different times for each sheet, and move at fractioanlly differnt speeds through the transport.

- Forms that look the same may be different as they are printed in different batches, possibly by different printers

- The operator and/or the scanner's autofeeder and transport can introduce skew into the image.

Different pick up times

Even if the physical paper is identical when images are scanned or received through the FAX they are never completely consistant and always get shifted slightly when they are loaded into the memory of the computer. That is, they are moved up or down by several pixels as well as lengthened, shortened, widened or shrunk slightly.

Different Printing

As noted previously, even forms that look the same to the naked eye may be laid out slightly differently due to differences in the printing process.

Skew

Skew can occur as part of the paper, as a complete image on the paper, or as individual fields within the form. When a form is loaded in a tractor feed or if it is placed at an angle in an autofeeder, the complete image can be skewed. If the form is completed with a typewriter, the data can be skewed. If the paper is a photocopy or fax, the image may be skewed.

Paper skew is caused by the rollers in a scanner's transport or if the paper is placed at an angle in the autofeeder. As rollers move at slightly different speeds all scanners cause a certain amount of skew in the image. Small papers that cannot be aligned against the guide bars may skew badly and sliding guide bars may get moved slightly wider than the paper. Batches of old and folded paper may not be cleanly jogged into tight alignment causing the guide bars to be set wider.

1971 Cadillac	**Cadillac**		
"75"Limousine (Fleetwood)	69733	D	8
1968 Cadillac			
"60 Special" Sedan (Fleetwood)	68069	C	8
Broughham (Fleetwood)	68169	C	8
Calais Coupe	68247	C	8
Calais Hardtop Sedan	68249	C	8
Coupe DeVille	68347	C	8
Hardtop Sedan DeVille	68349	C	8
DeVille Convertible	68367	C	8
Sedan DeVille	68369	C	8

Figure 10.2 Illustration of skewed paper and of skewed image on a straight paper

courtesy: TMSequoia Corporation

Deskew

Skewed images must be corrected. So before any software can start to locate the required data or the data to extract, the image must be deskewed and registered so that each page is located in a consistent place in the memory array.

There are two types of registration in use:

1. The form will consistently align based on the top left edge of the paper.

2. The form contains "registration marks", typically at least three right angled marks in the corners (to allow for triangulation), which the software uses to line up the form and stretch or compact it. Sometimes logos or other distinguishing marks can be used as well.

The latter allows forms processing to work more efficiently easier to implement, but it means that you must have some control over the forms design.

1971 Cadillac	**Cadillac**		
"75"Limousine (Fleetwood) 69733		D	8
1968 Cadillac			
"60 Special" Sedan (Fleetwood)	68069	C	8
Broughham (Fleetwood)	68169	C	8
Calais Coupe	68247	C	8
Calais Hardtop Sedan	68249	C	8
Coupe DeVille	68347	C	8
Hardtop Sedan DeVille	68349	C	8
DeVille Convertible	68367	C	8
Sedan DeVille	68369	C	8

Figure 10.3
After processing, no skew, backgrounds removed - size reduced from 53 KBytes to 33KBytes (see Figure 10.2 for original)

Image courtesy of TMS Sequoia

Skew causes problems with location and identification, but even if the software can accurately locate the fields, skew can interfere with accurate recognition. Most OCR packages can deal with up to 5% skew without a problem - but 5% is not the same as 5 degrees. Five percent skew allows a movement of 5 pixels per one hundred, which at 200 dots-per-inch resolutions equates to just a half inch out over an 8 1/2 inch page - a very small angle. Skew in forms data can be as much as 30 degrees which is 50% skew at 200 dpi!

A difference between forms processing and text processing is that text processing skew is fairly consistent across the whole image width;

but in forms processing it may vary in different areas of the form depending on the elements of the form and how the form was completed.

All scanners introduce some skew as some rollers pull slightly faster than others. The skew caused by the scanner's rollers may be hardly noticeable - 3-5 pixels across a line means that it is moving 3 or 5 pixels in 1,700 (1/300th of an inch) for a letter sized document.

Skew can also be introduced by the size of paper in an auto-feeder or by the operator. In these cases it may be substantial. As much as 30 degrees skew can occur on small forms when scanned in a mixed batch including larger ones. Sloppily placed forms scanned from a flatbed or manually fed can be badly skewed even to the point of losing information off the edge in some cases.

Deskew software will normally start by straightening the paper which occurs as part of registration.

Once the paper image is straight then the software can work on the image. This generally is carried out by taking the line along the bottom of characters and the lines of boxes, and then using those to correct the skew of the form within the page.

Deskew of the individual areas within the image requires extraction of the area and, because skewed text can cross over instructional text, needs additional processing to remove it (see forms removal later).

Grayscale and Color Deskew

Most deskew works on the bitonal image, but some scanner manufacturers and interface board vendors have now implemented grayscale deskew and color deskew will follow. Grayscale deskew is much more effective than bitonal as in some cases, pixels can get lost when they cross boundaries. The shading information carried in the grayscale or color information provides enough surrounding pixels to eliminate this. Grayscale and color deskew require the processing of up to 24 times as much information as bitonal and therefore is processor intensive and can slow down deskew.

Software Image Enhancement

Often the text or data information on a form, particularly that on a copy is damaged or faint (see figures 10.4 and 10.5). Alternatively the characters may have been created with a dot matrix printer. Another area that can cause damage is through lines delineating boxes interfering with it (see fig. 10.6), instructional text getting in the way, and bleed through or noise (a particular problem with FAX's). Image enhancement fills in the gaps in the characters, sharpens the edges of them, separates them from adjoining characters and removes noise. Some image enhancement occurs at the scanner level (see chapter on scanning) but is then supplemented with software. Enhancing the image in hardware on a bitonal scanner is attractive when making background differentiations as the scanner has reference to all the tonal information before it is converted into bitonal (i.e. black and white).

Figure 10.4
Example of
carbonless
print.

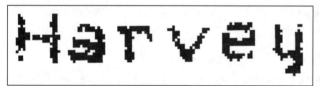

Figure 10.5
Enlargement of
"Harvey" in figure 10.4
showing break-up in
characters

Image enhancement software utilizes the bitonal image to further fill gaps and sharpen the characters. It works on the basis that a character or a line always consists of more than one pixel or dot. Don't forget that we are scanning at 200 or 300 dots-per-inch. Therefore any single or possibly dual white pixel gaps surrounded by or connected to black pixels can be filled in, and any single black pixels surrounded by white space are probably noise. When setting filters to add or subtract pixels however, the user must be careful not to remove small portions of information that are important to the recognition algorithms that follow such as the dots over an "i" or 'j' or very thin lines that are part of a barcode.

Forms Removal

Having straightened the form and the image, removed extraneous backgrounds and sharpened the contents it is now necessary to recognize and extract the information that is required in the computer and jettison all the rest. Forms contain much information that is irrelevant to the capture process. In fact if you look objectively at a form you will see that the amount of data that is required to be extracted from any one form is invariably a fraction of the total of what is there.

Forms carry out many functions as well as a method to transfer information.

- Forms are used partly as a marketing tool, promoting a company's look - after all they are sent to all a company's valuable customers. The look of the form tells the user something about the care and attention to detail of the company and the image it wants to portray.

- A form is partly a legal document. It contains language that protects the company from litigation and spells out the terms of doing business.

- Finally a form has to be understandable to a person who may never have seen it before, so it contains all sorts of instructions.

Important though it is to the sending company, all this type of information is irrelevant to the information or data that is required for processing.

Examples of this irrelevant information includes logos and artwork, name and address of the sending company, locational boxes, instructional text at the bottom or top of the boxes, legal disclaimer or non-disclaimer notices, forms identification numbers and general information. Either all this data must be removed from the form before recognition is commenced, or alternatively the required data must be extracted from the form.

So most forms are printed in bright colors. One of these is red which had a big advantage with previous generation scanners that used tungsten bulbs which were blind to red. Unfortunately today's scanners tend to use extra bright fluorescent bulbs or LED's which are normally green. These types of bulbs convert red to black. So the red forms are displayed as badly as any other bright color. To solve this various manufacturers have provided "red light" versions of their scanners, which, through the use of red fluorescent bulbs are blind to red. Of course this is unselective - all red goes regardless of the source, so you cannot easily use one of these scanners as a general purpose scanner -- a red check mark for instance will disappear! Some scanners are tuned specifically for red drop-out, others permit the operator to change bulbs and a third group incorporate both light sources allowing the operator to switch using the control panel.

There are two ways to remove forms which do not use drop-out inks.

The first, which is the simplest and can be used for unknown forms, just looks for straight horizontal and vertical lines of pixels and removes them. The instructional text, logos, names and addresses and boilerplate legal information etc. all remains. To solve this problem the software in some systems will assume that anything of say 9 point type is instructional text and should be removed. This method allows some reduction in image storage sizes and improves recognition of the data you want. The problem is that unless a manual step is carried out to visually locate and clip the extraneous information, any recognition carried out in the next step is going to recognize everything that remains.

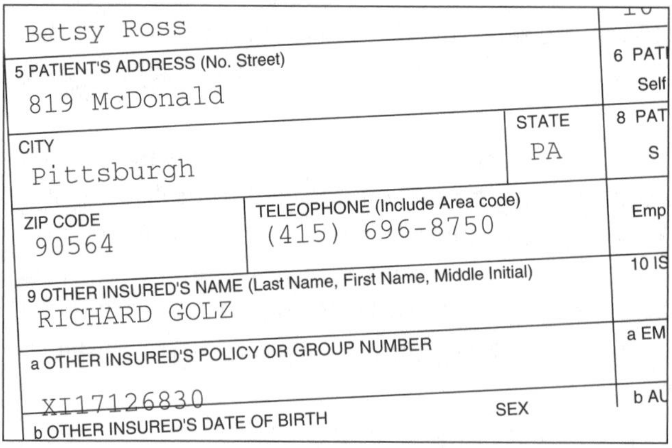

Betsy Ross
5 PATIENT'S ADDRESS (No. Street)
819 McDonald
CITY
Pittsburgh
STATE
PA
ZIP CODE
90564
TELEOPHONE (Include Area code)
(415) 696-8750
9 OTHER INSURED'S NAME (Last Name, First Name, Middle Initial)
RICHARD GOLZ
a OTHER INSURED'S POLICY OR GROUP NUMBER
XI17126830
b OTHER INSURED'S DATE OF BIRTH
SEX
6 PATI
Self
8 PAT
S
Emp
10 IS
a EM
b AU

Figure 10.6 Part of a Completed Medicare form. Lines, instructional text and skew interfere with OCR capture Policy number crosses line below

(courtesy TMSSequoia)

```
Betsy Ross                      10

819 McDonald

Pittsburgh                      PA

90564          (415) 696-8750

RICHARD GOLZ

XI17126830
```

Figure 10.7
After form Removal,
ready for OCR. Removal
of background data
leaves clean text for
capture. Characters
are repaired after
line removal

(courtesy TMSSequoia)

The more sophisticated systems use a blank template of the form to identify the form, line up the completed one with it and then decide on the elements to remove (see figure 10.7). In this case, a user obviously requires access to a blank form. If the form does not have anchor points then template removal software starts by choosing an identification area - typically a distinctive pattern such as a logo or area that is easy to locate and in a constant location. This represents the "anchor" point that is used to identify the form and position it so that the correct template can be retrieved and applied.

A problem that can occur on computer or typewriter completed forms, but that does not occur on electronically created forms, is that sometimes required data crosses the lines or instructional text. This can happen due to wrong positioning of the form, bad design of the form or may be deliberate as with a rubber endorsement stamp (see "OTHER INSURED'S POLICY OR GROUP NUMBER" field in figure 10.6.)

In these cases recognition problems can be introduced, unless the system can repair the characters. This is carried out either by keeping the parts of the line that may damage the character if in doubt, or by adding back any removed pixels based on an expectation of what the character should look like (see the section on software enhancement above).

Some systems will let the user scan a completed form and then draw around the data areas to extract using a mouse. The advantage of this type of product is that you can utilize a received form from say a supplier as a template without needing to call him and get a blank form. And it is fine if the boxes are fairly well delineated with the data easily bounded by the boxes. But if the data boxes are narrowly defined or a lot of data is on the form, then it is sometimes difficult for this type of product to identify what is data and what is background.

Another problem that can occur when forms boxes are placed very close to one another is that the system can get confused and not be able to accurately pinpoint the data that goes with a particular zone. As a result the last few characters of one zone can get picked up as the first two characters of the next zone. This can occur particularly on forms with multiple lines such as purchase orders, invoices, or charge lines on insurance claims and when data on the form is skewed across more than one box.

Data Extraction and Formatting

Once the form is removed, the image consists of all the data that is needed.

In the case of dropped out forms, a template is maintained for viewing and printing purposes. The image can then be superimposed on the template.

In the case of forms removal, the system will normally maintain an original image, complete with all the forms information for viewing etc..

This must then be separated into different types of data - printed, optical marks, barcodes, fields that must be manually keyed etc. and those that can be passed to the appropriate recognition engine(s). Alternatively, if the data was tagged and extracted, then normally one ends up with many small image files each containing extracted areas of image sometimes known as "snippets". Each of these miniature image files usually contains the name of the original image so that it can be identified and put back together as a data file.

Critical fields that need to be 100% accurate can be tagged for recognition and subsequent key verification.

CHAPTER

Character Recognition

Two different terms are usually used within forms processing to identify character or text recognition, OCR and ICR. Both OCR and ICR are methods of converting visually readable characters into computer readable characters. The differences lie in the methods used and therefore the types of characters recognized.

Many people start with OCR thinking that it will accurately convert the text data if all the pre-process is carried out correctly. The first thing to realize is that it won't and the second thing to realize is that ANY LEVEL OF RECOGNITION, when combined with the right post-process tools will reduce labor costs.

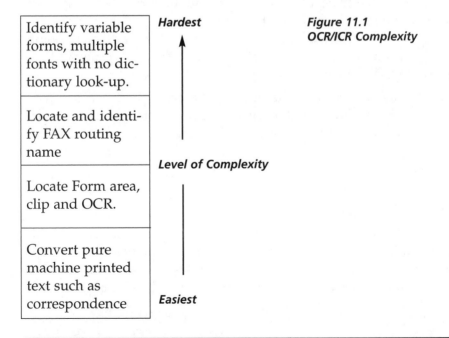

Identify variable forms, multiple fonts with no dictionary look-up.	**Hardest**
Locate and identify FAX routing name	
Locate Form area, clip and OCR.	**Level of Complexity**
Convert pure machine printed text such as correspondence	**Easiest**

Figure 11.1
OCR/ICR Complexity

OCR for most forms processing helps substantially and even a low percentage of recognition can be cost justified. But it is important to understand that it will not recognize everything accurately. The chances of success depend largely on the original media

OCR, which stands for Optical Character Recognition, is used to identify the software that can recognize printed standard fonted characters regardless of their font size

ICR which stands for Intelligent Character Recognition is a trainable software system which is used to identify hand print and fixed font characters such as OCR-A (see fig 11.3) or OCR-B (see fig 11.4) or Magnetic Ink Characters (MICR) such as E-13B (see fig 11.2).

Figure 11.2

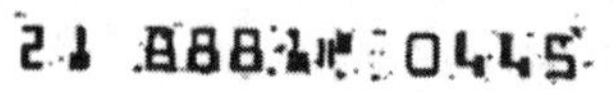

2⌒ ⸖888⸗1⸗ ⸖0⸗44⸗5⸗

Example MICR Line "E13B"

Some manufacturer's recognition engines work well with some specific fonts or types of text while other engines may work better with other fonts or text. So the user may need to use different products depending on the types of data contained in the form. Some types of data are more reliable in certain circumstances than others. For example: most OCR works well on typed characters that are 10 or 12 point. Some OCR is more tolerant of very large fonts such as you would get in a **HEADING** than others.

ICR works well with hand printed characters and because it is usually neural net based and it can also be trained to learn other characteristics such as when an optical mark is present (see next chapter).

Some vendors of OCR allow on-line edit verification, which can remove a post-process function, but it slows down the process and is not suitable for high volume forms processing.

The output from these engines also needs to be considered, some is easy but others need interpretation. If it is text that is converted with OCR or ICR then that is fairly easy and the separate fields can be separated or delimited with commas. Barcodes likewise will create a series of numeric or alpha/numeric characters. Check marks though need to produce interpreted code.

CHAPTER ELEVEN

OCR

OCR came first. It started with conversion of a highly stylized font known as OCR-A, that was difficult to read by humans. OCR-B followed with a font that was a lot more readable.

OCRA 123456789

123 456 789

Figure 11.3 Example of OCR-A

OCRB 123456789

123456789

Figure 11.4 Example of OCR-B

Omnifont OCR, originally designed to save re-typing textual documents like correspondence, appeared in the 1980's. ICR, which originally was used by Kurzweil (a division of Xerox) to denote their superior OCR recognition arrived later. Nowadays ICR is often used as a term to denote handprint recognition.

OCR-B is still used on checks and other documents in parts of Europe, and neural net OCR is used to supplement the MICR (Magnetic Ink Character Recognition) in the US and Britain (E-13B font), and France (CMC-7 font). In these cases, the font size and shape is very specific so it can be specifically learned.

Omnifont recognition has been used for text documents for a number of years in low volume environments and forms processing vendors have used the engines to automate the capture of the data from the forms, but this creates challenges.

OCR looks at the shape of the bitmapped image of each character in turn. Then it compares the shape to preset patterns and converts them to the ASCII equivalent code. This is usually supplemented with dictionary look-ups to ensure reasonableness. OCR and ICR engines all provide a likliehood percentage. For example in the figure below, the engine might say that there iis a 90% chance that the character (which is well formed) is a 2. The next most likely character may be an 8 with perhaps a 5% likliehood. These percentages are used by forms processing systems in conjunction with other checks such as look-up tables, reasonableness, edit checks and sometimes the results from other engines to increase accuracy.

There are two major ways that OCR uses to decide on what a character is: scoring using a template mask and pattern matching.

Scoring:

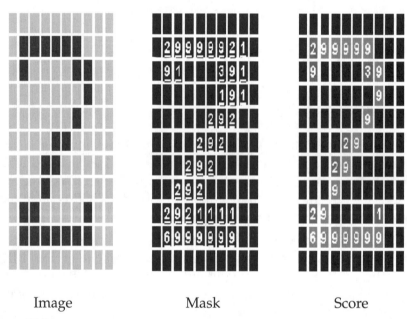

Image   Mask   Score

Figure 11.5
Example Image of character matched to Image Mask to create scored image
(courtesy Banctec)

Under this method, the characters are matched to pre-defined templates of all the characters and a mathematical formula is used to assess what the character is most likely to be.

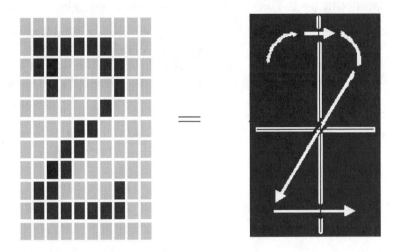

Figure 11.6
Feature Analysis sometimes divides the character into quadrants.

In Figure 11.6 above in simple terms the engine might look for characteristics as follows:

the top left quadrant has a curve up and a straight line,

the top right quadrant has a straight line, a curve down and a diagonal line from right to left,

the bottom right has a line,

the bottom left has a diagonal from left to right and a straight line

As with pattern matching it will score how close the match is and provide a perventage likelihood with an alternative.

Neural Networks pass fonts through a learning process. The pattern of each character is learned by the system and variations are learned through repeated training.

Most engines mix the technologies to enhance overall recognition providing pre-set learned fonts to minimize learning. However all OCR engines have a preference - that is one vendor's product works better with one form of text than another. Some lower end systems tend to use one OCR engine, but the higher end vendors have implemented the use of more than one. Some vendors even identify the type of font and try to direct it to the most effective engine.

Another way to solve this issue, is to use so called "voting engines" which will pass the data across multiple recognition engines (normally three different ones) and "vote' for the most likely result. This has been shown to increase accuracy of recognition by 50% or more on problem text.

Figure 11.7 Voting Example

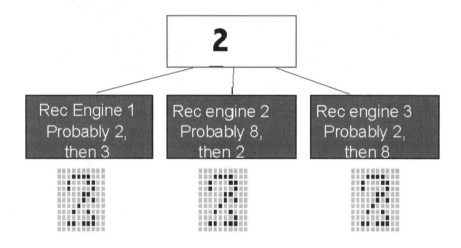

Voting Example.

Voting uses multiple engines to receive a result and then utilizes the percentage likelihood of a character to decide which character it is.

It's easy to locate a form identification number, clip the area from the bitmap and pass it to OCR for an automatic index. It's fairly simple to look at a FAX cover page and identify a routing name in printed characters. Forms however contain different fonts sometimes created by different systems and character strings, such as invoice numbers, cannot be dictionary verified.

Even though recognition can recognize many of the characters, there are a number of other issues specific to forms processing which reduces accuracy. Most OCR relies on dictionary look-ups to provide contextual accuracy, but the problem with forms information is that it is often not words that a dictionary would recognize - such as names and addresses, account numbers or part numbers and descriptions or codes.

Assessing Quality

Good quality text on simple forms can achieve over 99% accuracy, but poor quality can lower rates to as low as 50%. When OCR accuracy is measured however, it can only include errors that it knows it made, unknown characters or characters it is not sure about. These are what the statistics measure. Anyone can achieve a reported 100% accuracy by simply accepting those characters that the engine reports back with say over a 50% confidence factor — but this will create a bigger problem.

This problem lies with substituted characters - those that the OCR recognition engine is convinced that it got right and that passed its reasonability checks, but which in fact are wrong. I describe these errors as "unknown unknowns". An example might be the word OATS. If part of the O is mssing then the OCR might convert it to CATS and not be any the wiser, but someone receiving a bagful of CATS on his invoice might not be too pleased. Substitutions when added to errors increase the true level of errors substantially. Often they are caught through edit checks and validations, but some can only be caught with a key verification step.

An additional problem, which we will address in the 'repair chapter', lies in character errors vs field level errors

However even a 50% accuracy rate in the forms processing market can be attractive. Consider a low volume user scanning 5,000 forms a day and the average form contains 100 characters of data to capture. This gives a total of 500,000 characters a day to capture. In this case a 50% recognition rate will convert 250,000 characters automatically. At an average key entry rate of 10,000 keystrokes per hour and a fully loaded labor rate of say $20 per hour, the user will save $500 a day from key entry alone.

Resolution

There is a popular misconception that to get accurate OCR, you must scan at 300 dpi resolution. It is true that In some cases, OCR products have been tuned to a 300 dpi size which gives over twice as many pixels as 200 dpi. However with most modern OCR engines it doesn't matter as much as insuring that the image is clean, sharply delineated and straight. These are issues mostly associated with the quality of the scanner and the amount of pre-process. Having said that, if fonts are so small that they can not be resolved without using a higher resolution, then OCR is not going to work unless you scan at a higher resolution.

Any suspect characters must be labeled and the appropriate part of the image (known as a zone or snippet) must be clipped or identified via X,Y coordinates for subsequent repair.

ICR

ICR (Intelligent Character Recognition)

I differentiate OCR from ICR loosely, by thinking in terms of OCR to convert machine printed text and ICR to convert handprint, sometimes optical marks, and also for specific fonts, or as a supplement to OCR.

Handprinted characters vary in shape, creation and placement, so recognition accuracy must be improved through constraints including:

- Limiting the character set (for instance numbers only).
- Limiting the characters to upper case only.
- Confining characters to boxes.
- Utilizing specific look-up tables - for instance: employee names for FAX routing.

Hand writing is generally categorized in three different ways which get progressively more complex to recognize. These are known as:

- Constrained Handprint (letters are placed inside delineated boxes);

Figure 11.8
Constrained handprint

- Unconstrained handprint (letters are printed - not connected - but they may be placed in different spaces and there are no constraining boxes);

Figure 11.9
Unconstrained
handprint

- Cursive Writing.

Figure 11.10
Cursive writing

Forms with constrained handprint on them were generally designed specifically for automated recognition. The handprint is constrained by asking the person filling out the form to write inside boxes - which are normally faintly outlined with a drop-out ink. Constrained hand-print can be recognized very accurately with today's tools and is the preferable solution.

- Constrained handprint example (the shaded boxes force the user to print within them) see following page.

- Unconstrained handprint requires the forms user to have some measure of control over the person filling out the form as they must complete it neatly and not connect the characters for good recognition. However, the terms "print clearly and legibly" are usually aimed at getting a person to print the characters so that they can be recognized automatically. Unconstrained print is adequate to be useful if the characters are limited but recognizing freeform printed writing such as found in a letter is still in early stages. Clearly printed text is capable of around 50-60% recognition accuracy at the time of writing.

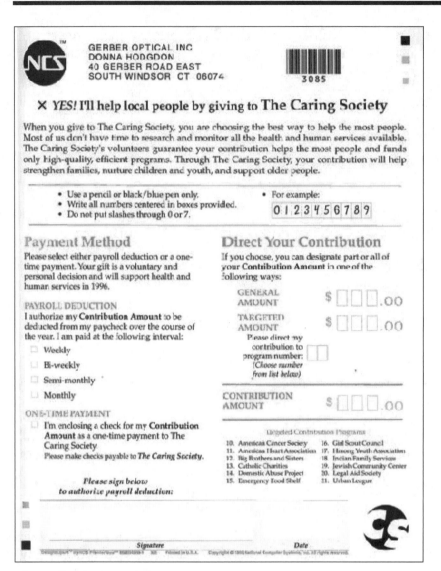

Figure 11.11
form with drop-out inks for constrained hand-print

- Cursive recognition of mixed alphabetic/numeric handwriting does not work well at all and has a long way to go. But if the character set is strictly limited it can be successful. For example banks have had good success with Courtesy Amount Recognition (i.e. recognizing the amount on the check) which is known as CAR by knowing that each figure must be between the numbers zero and nine. Recently this has been supplemented with Legal Amount Recognition (LAR) which reads the written amounts and compares with the CAR amount.

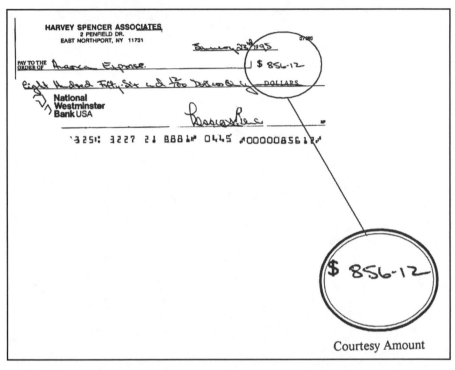

Courtesy Amount

Figure 11.12
Courtesy Amount recognition converts dollar numbers to computer date

As with OCR connecting different recognition methods together can improve the recognition levels. For instance some users have found that handprint recognition works best by combining feature based extraction techniques with neural network recognition.

In all recognition any suspect characters must be labeled and the appropriate part of the image related to the suspect character (known as a zone or snippet) must be clipped or identified via X,Y coordinates from the bitmapped image. This is then sent to the repair system with the 'mostly likely character' for subsequent operator viewing and manual entry.

Barcodes & Optical Mark Recognition

Barcode Recognition and Glyph Recognition

Barcodes can provide an extremely reliable method for capturing information from a form, and the barcode can be found and decoded even if other data or noise is interfering with it. But someone has to create the bar or stick it on the form.

Barcodes come in various defined formats that are used in different industries. We're most familiar with the UPC code used in supermarkets and accept the barcode used to correctly route mail as part of the envelope. But there are a number of other formats used known by various acronyms: EAN (the European version of UPC), 3of9 also known as Code 39 a common code as it was the first alphanumeric code developed, Code 128, Code2of5, interleaved 2of5 and a version of that used by the airlines at the bottom of the tickets, Codabar used by FedEx etc..

All these codes are linear one dimension barcodes i.e. the code line spreads across the bars. Recently we have seen the introduction of a number of new two-dimensional barcodes. Two of these have been discussed and used for forms processing. One is named PDF417 from Symbol Technologies and the other Smartpaper from Xerox. These two codes are fundamentally different, with PDF-417 being in the open domain and more conventional while Smartpaper glyphs have remained as Xerox proprietary and therefore have not been used much.

Figure 12.1 Sample Barcode

Barcodes have a defined format. In the case of the standard barcodes a certain number of thick bars combined with a certain number of thin bars makes a character. It's a similar principle to morse code which uses dots and dashes. The Smartpaper from Xerox works on a different principle with diagonal slashes representing binary ones or zeroes.

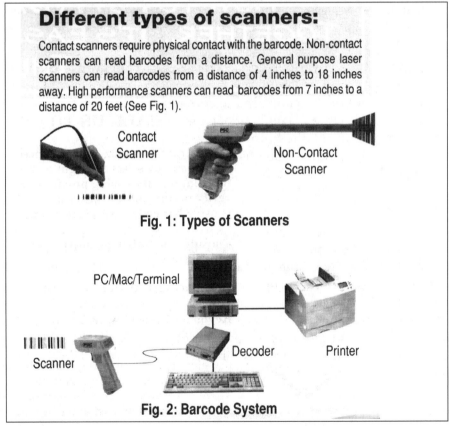

Different types of scanners:

Contact scanners require physical contact with the barcode. Non-contact scanners can read barcodes from a distance. General purpose laser scanners can read barcodes from a distance of 4 inches to 18 inches away. High performance scanners can read barcodes from 7 inches to a distance of 20 feet (See Fig. 1).

Contact Scanner

Non-Contact Scanner

Fig. 1: Types of Scanners

PC/Mac/Terminal

Scanner

Decoder

Printer

Fig. 2: Barcode System

Figure 12.2
Courtesy Systems IDWarehouse

These are known as Glyphs. Normally users print the human interpretation underneath the barcode but this is not essential.

Barcodes in most applications are recognized with a traditional laser scanner (like the ones in the supermarket) which is attached to the scanner or with a wand. In document imaging applications, batch control cards or bar-coded folders may be 'zapped' with a hand held unit or the barcodes may be identified from the bitmap image. Some scanners offer a barcode reader within the scanner. This is typically a laser based scanner that the document moves underneath. The advantage is that the barcode can be very dense but the disadvantage is that it must be located in line with where the scanner is located.

In forms processing applications, barcodes normally form a part of the data - usually a control number or document id number. Often used in the transportation industry, a typical example is the Federal Express waybill with a barcode denoting the shipment number coded as a barcode and printed clearly underneath.

Bar codes placed with sticky labels on forms are being used to access and track forms or identify the type of form. Some sites are using OCR for this purpose but the results can be unsatisfactory as the document's are often in poor condition, and the type and size of font used are not easy for OCR. As a result accurate conversion rates are frequently less than 90%. Even if the user can achieve the 98-99% accuracy claimed by manufacturers, a manual verification process must usually occur. Bar codes however, can be converted with nearly 100% accuracy from the image once they are located. You can guarantee that you have the whole field since it is bounded by start and stop bars. The defined check digits can ensure that you have correct data.

Depending on whether the form has gone through a pre-recognition process to identify and extract relevant data, there may be one or two parts involved in reading and decoding image bar codes - location of it and decoding it. If the form is internally generated then predefined barcodes can be positioned at specific locations for the software to decode. However, if the user has no control, the bar code may be located anywhere on the document and the formats may vary. Typically a person sticking on labels will be instructed to stick it anywhere where there is a gap so as not to obscure any important information. It can then be located using pattern recognition (bar codes have a very distinctive black and white pattern). But there is a time penalty. It is substantially quicker to decode it if the software knows where to look in a small location or if it is in a snippet, than to look all over the form.

The FedEx waybill has its barcode in a fixed location. By providing approximate coordinates to a program, location can occur in a fraction of a second. After location - recognition - takes over. FedEx's Codabar scanned at 300 dpi for example, has wide bars and spaces approximately 14-16 pixels wide and thin ones approximately 5-7 pixels wide. It is comparatively simple for a program to be written to decode this information reliably. Skew in the paper or the barcode is much less of a factor with barcodes than OCR - generally barcodes need not be deskewed as long as the software can effectively draw a straight line through the bar and see the differentiating black and white pixels.

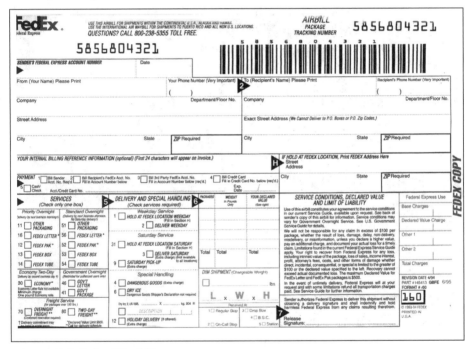

Figure 12.3 Fedex Waybill

Currently most formats apart from UPC are available from a number of manufacturers. Some integrators have added laser barcode readers to a scanner's front-end as a simple integration. For batch card identification this is fine since it assumes that the barcode is located within a line of the reader's location and it allows for fine barcodes to be used. This reduces the space required, but for identification of individual documents it may be less satisfactory if the bar cannot be always placed in the same location. Otherwise, it is available through scanner co-processor vendors or through software.

Although most barcodes are easy to read and implement, there are still guidelines to follow to achieve maximum efficiency when using software. Not all barcodes are created equal as different barcode types have been designed with differing objectives in mind.

Most barcodes consist of a fixed number of bars per character and many have a defined check digit to enable verification of the decode. If a bar is missed, the barcode becomes unreadable. If adjacent bars are touching, it is possible that the barcode's size is misread, but because the table of bars is smaller than the number of possible combinations, an illegal character will be generated and the decode rejected.

A barcode is thus significantly more secure than OCR since substitution is virtually nonexistent.

The most popular barcodes in imaging are

- Code 39,
- Code 2of5
- Code 128
- Codabar

The following is a description of each with the advantages and disadvantages.

Barcode types differ with the encoding system used. The binary coding systems, Code 39 and Codabar, use only thin or thick bars so it is simple to convert these into numbers. Code 128 is a proportional code that uses different sizes of bars, with the size and sequence defining the character represented. This type of code is more difficult to read and requires more accuracy in printing. The advantage however is that more data can be coded (alphanumeric, numeric, special characters).

When deciding whether to use barcodes and what formats to use, the user should ask himself seven questions:

1. What character set needs to be encoded (alphanumeric, numeric, special characters)

2. How much data needs to be encoded within a given area. Barcodes are intrusive but some codes are more compact allowing more data to be encoded or less space to be used.

3. How much room is available on the document(s). Make sure you include space for the "quiet zone" which is a space at both ends of the bar. This space is critical for locating and good decode. For example, the Fedex waybill using Codabar takes up three-and-three-quarter inches in length to encode just 12 characters.

4. What scanning resolution is available. If you are intending scanning at 200dpi the type and size of barcode is significantly reduced.

5. What is the degree of tolerance to misreads. If a barcode fails, the data must be entered via keyboard. The user must ensure that a correction mechanism exists within his application to perform image assisted data entry of unrecognized documents.

6. Does the complexity of the data require a check digit or parsing algorithm to separate sections of the barcode into different index fields.

7. Possible position on the document. Certain barcode types are more tolerant than others to being located on the edge of a document for example.

Basic rules:

- Users should always select the simplest and largest barcode that their application will allow.
- Leave a minimum "Quiet Zone" or border of 1/4", preferably more, around it.
- Always print the translation above or below the barcode to ensure readability for key entry or OCR validation.
- Use a check digit such as Mod 7.
- Create a large height to width ratio to allow multiple seeks (at least 1/4")
- plan on scanning at the highest resolution possible.

Resolution and Density Issues

Unlike what the barcode industry and laser based barcode scanners require, the user of forms processing systems must concern himself with the scan resolution. Obviously, the higher the scan resolution the higher the barcode density may be, although the user must ensure that the bars are not so close together that they touch. Poor print quality from a dot matrix printer or highly skewed images will increase the probability of lack of scan separation between the bars.

Most imaging systems scan at 200 or 300 dots per inch although standard faxes have a resolution of approximately 100 dots-per-inch on the vertical plane. The scanning resolution combined with the needs of the decode software affects the size of the barcode by restricting the density and therefore the amount of information held on the barcode.

Different manufacturers require different numbers of minimum pixels to decode accurately, some require 2 while others require 4.

Barcode specs are usually defined in mils (i.e. .001 inches) so this means that at a 200 dpi resolution scan will require the thinnest bars to be printed between 10 mils and 20 mils thick (see also the section on barcode printing below).

Figure 12.4
Illustration of various bar code symbologies

Before selecting a barcode and density find out which the software requires. Overall though, a good rule of thumb is to allow 4 characters/inch at 200 dpi and 6 characters per inch at 300 dpi.

Glyphs are a different matter as they consist of a mix of '/' and '\' characters. When these are printed densely on the page they form an attractive halftone pattern that the user can printed on top of (see figures 12.6 and 12.7)

Locating the barcode.

In fixed position; floating position, angled, upside down or sideways.

Some software products have hard coded the search location of the bar code. This technique, while perhaps adequate for predefined batch header sheets and preprinted barcodes, lacks flexibility. Even if the barcode is always located in the same place, a page may have been inadvertently placed upside down within a stack.

Now though, most vendors allow users to either find the barcode wherever it is or to specify multiple approximate locations to find the barcode(s). These may be mixed types and different numbers on one page. Locations may vary as the barcode may have been put on the paper after it was completed, positioned to ensure that information isn't obscured. As these types of barcode may be stuck on rapidly, they may be at acute angles.

In transportation and inventory control type applications, the barcode label is also pasted somewhere on the form but neither location nor angle is known. A typical strategy implemented by software vendors is to initiate a semi-sequential seek to search every 1/4" until a barcode pattern is recognized. As this is time consuming, it is preferable to know approximately where the bar code is located rather than let the software find it.

The angle of tolerance is determined by drawing a horizontal line between the upper left and lower right hand corners of the barcode. If the angle exceeds the tolerance, the software usually initiates another search from the opposite direction to locate the vertically positioned bars. To catch the labels that were pasted on at angles outside the tolerance, the software must search at an oblique angle, increasing the processing time significantly. Because of this it is preferable to print a location box on the form if possible.

Deskewing and Image Enhancement

Deskewing barcodes is not a difficult task as the software knows it has to create straight lines. Unfortunately though some image enhancement hardware or software can bleed together barcoded lines when it is trying to sharpen text. The only solution to this is: either switch off the image enhancement filters; or capture the grayscale and process the barcode separately as lines instead of curves.

Processing Time

Barcodes are simple codes that can be decoded very quickly. Depending on the number of seeks and whether the software uses a co-processor board, or the main CPU, single barcode decodes can be performed in anything from a half to a small fraction of a second.

Integration of Barcodes into the System

Despite the simplicity of barcode reading, not many people have implemented them.

It is easier for people to accept OCR as it is converting human read-able data and barcodes can be intrusive and are not very aesthetic. Also, many people look at a barcode and consider it complex because they have not dealt with them. But barcodes are simple to use, fast, reliable, and are not subject to decode errors from noise or other marks nearly as often as character based OCR data. They can be quickly located with software, are not subject to the image degradation and many codes include check digits and self checking symbologies..

Barcodes are popular for proof of delivery applications, inventory control, and in transportation. In addition, FAX'd barcodes can be accurately interpreted - often a problem with text and OCR. However a major problem with barcodes is the amount of space needed to put them on a packed form.

PDF 417 is capable of storing a large quantity of information, at ,0075 inches (7.5 mils) which is about 130 dpi, this code can encode up to 500 characters per square inch which makes it a good candidate for storing complete forms information. It can also be extremely secure—checking the data decode and correcting for erasures (missing or undecodable characters). For example at the highest security level, up to 510 codewords, which equates to 612 bytes of information, can be obliterated with a hole or other mark without losing any data at all.

But there is obviously an overhead in the space that is required. As an example it will drop to below 300 bytes/sq inch to achieve a 27% error correction.

Figure 12.5
PDF 417 example-
simple HCFA Form

Figure 12.5A
HCFA with PDF Encoding

Smart Paper Glyphs are another alternative. At 300 dpi 398 bytes can be encoded per square inch, which drops to 298 with 25% error correction. The advantage of data glyphs however is that printed information can be placed on top of them without impacting the decode capability. The glyphs look like half toning which can be incorporated as part of the forms design.

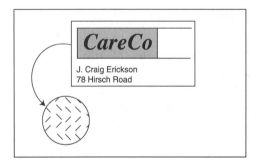

Figure 12.6
Example of glyphs in use as a shaded background behind the words CareCo.

Courtesy of Xerox Corp.

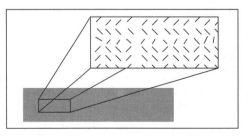

Figure 12.7
Smart Paper Diagonal lines known as data glyphs represent binary 0's and 1's when placed compactly together they form a shading

Courtesy of Xerox Corp.

The problem is that if they are printed densely, the scanner's thresholding will attempt to remove them as background. But if they are printed larger then they become intrusive.

Barcode Printing

The quality of the printing and the paper quality affects the performance of a barcode reading system. On a form a preprinted barcode that was printed when the form was originally generated offers the best performance, but if labels are stuck on during batching then it is preferable to utilize a standard thermal printing device. Avoid colored inks - black is best, and ensure that the glue on any label is very firm so that the rollers on a scanner do not peel it off. Experts also advise that you use high quality materials as lower quality can damage the thermal head.

Laser printers are the next best choice, providing excellent contrast and clean separation between bars, followed by ink jet, while dot matrix is affected by the quality of ribbon. Users creating barcodes with dot matrix must constantly monitor the quality.

Do not make the thin bars too thin - most software requires two to four pixels across to ensure accurate decode and each bar must be separated by at least one and preferably several pixels. A number of factors have a bearing on this distance. First the density of the barcode itself. The denser the barcode, the less distance between the vertical bars.

Specialized scanners using laser beams have an advantage since they can capture at very high resolutions of up to 1,000 dpi. Certain barcodes tolerate densities of 10, 12 or even 18 characters per inch, but each barcode requires a minimum number of pixels to decode accurately. For example Codabar requires 32, Code 39 requires 50. This means the maximum density of a Code 39 barcode scanned at 300 dpi cannot exceed 6 characters per inch and 4 characters per inch at 200 dpi.

Optical Mark Recognition

Optical mark or mark sense recognition has always been a popular way to get people to complete forms. Mark sense was a system developed with punched cards to enable automatic data capture in the 1960's. It was based on electrically sensing the magnetic conductivity from a mark made by a soft pencil. This has been replaced by optical mark recognition

Optical mark advantages are that:

- they are easy to understand.
- they are quick to complete.
- they are fairly easy to decode accurately.
- the answers are easy for both machines and humans to understand without interpretation.

The disadvantages are that optical mark forms are not particularly user friendly, they are good for yes / no type answers but do not lend themselves to text answers or capturing names and addresses.

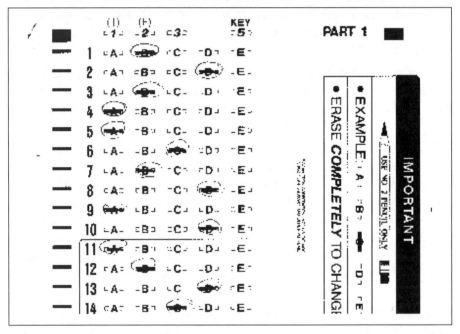

Figure 12.8
Typical OMR Form as used in an education test

The OMR forms used in education and surveys as shown in figure 12.8 contain timing marks down the side of the form. These are the thin black lines down the left edge of the form. These are used by several proprietary OMR scanners to insure that the correct line is interpreted.

Checkbox recognition in forms processing is sometimes known as OMR as it uses the same pattern recognition that is used for true OMR forms.

CHAPTER TWELVE

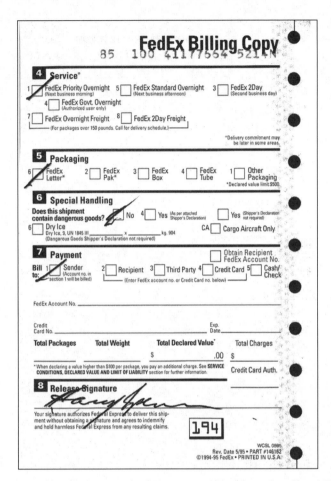

Figure 12.9
Simple OMR check off

Check-boxes vary dramatically in decode complexity depending on the design of the check boxes and the incentive for the person completing it. One extreme is multiple choice questions - if a consumer does not fill in the multiple choice questions circles completely, - he does not get the prize which represents a BIG incentive to do it properly and carefully. At the other extreme is answering the check-off boxes on a trade magazine's "Free Product Information" at a trade show. The boxes or circles are usually close together and a user has limited incentive to carefully fill in the right boxes without crossing over the next one. Sometimes they will even put a straight line down all the boxes.

Decoding:

The easiest and most common usage of optical mark recognition in forms are those used by education establishments. The student completes a series of bubbles with a soft pencil. The completed form is then run through a specific optical mark scanner that uses special light to measure the reflectance of the paper and the marks. These applications are dense and so need special paper and careful printing to ensure correct location of the boxes.

Forms processing systems using a bitmapped image can dramatically reduce the cost of printing as the forms can be copied on a photocopier or produced on a laser printer. In addition they can add value with extra recognition such as OCR'ing the name and address of the student. However the scanners for document imaging tend to be more expensive than dedicated low end OMR scanners so many people have stayed with OMR products despite their restrictions.

In forms processing, sophistication varies dramatically. Most forms processing vendors make a pixel count, letting the integrator/user set a confidence threshold. For example it counts the pixels that make an empty box, then it assumes that if that changes it means someone has made a mark.

Some use a neural net which using pattern matching to look at the pattern before the mark was made and the pattern afterwards. ICR, which is neural net based, can therefore be used in this way to convert check marks to data.

If the boxes are more than a certain percentage over the blank form, they assume it is completed. If you change your mind and scrub out the original or if your line accidentally crosses another box as you sweep your line elegantly across the box it may get recognized. To solve this problem forms processing vendors have come up with an upper or lower confidence level. E.g.. if the box is over 20% full and less than 80% it is accepted. This, together with rules to determine that only one box can be completed, works effectively in most cases. However, many will argue that to guarantee accuracy you need grayscale.

Other techniques that I have seen include:

- looking at the beginning of a line and at the end and extrapolating the center where the most likely box being checked is located.

- Identifying the types of marks that someone makes (e.g. X's) and looking for those. This is dangerous as sometimes one person will partially answer the questions and then pass the form to a companion or subordinate to complete. As more than one person will have filled in the answers they will be marked differently.

Edit checks

Most OMR lines only allow the person answering the questions to pick one box. For example consider the survey that asks you to fill in one of five boxes to express your satisfaction in a product or service.

1= dissatisfied,

2 = partly dissatisfied,

3 = neutral,

4 = mostly satisfied,

5 = delighted.

A user can only fill in one box, so the software will search across and accept the most likely, discarding any that are completely empty and any that are slightly full.

Decode output

OMR output needs to be converted into meaningful data. Two boxes may be Yes or No, in which case the software needs to identify the right box and output the result. Alternatively, using the above 'satisfaction' survey, the software has to convert the checked box to a number one through five. Often multiple different output conversions may occur on one form so the "instruction" tables to the software must identify the conversion required.

Another type of check mark which involves multiple data types is the box that is checked off to denote that the person completing the form wants to write in a comment. An example of this would be a change of address on a remittance. In this case the comment can be snipped out as a sub-image and sent for separate processing.

Other types of data

Sometimes forms contain other types of data such as lines crossing a continuum or logos.

Continuum lines are used in surveys or medical records to note where someone feels within a range. Typically there is a straight line labeled "poor to excellent". A forms user will cross that line at a point to represent what they "judge" about that item. Imaging based software can decode this type of data easily, applying a number to the location on the line based on neural net analysis. However it is not used much and I have only ever seen it implemented once.

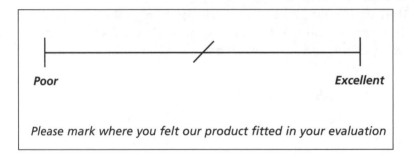

Figure 12.10 Continuum Chart

A logo may be interpreted using pattern recognition or may be used merely to identify a form type. Any neural network has the ability to recognize a logo and apply a value to it.

CHAPTER 13

Post Recognition Processes

Validation

As a human, when you interpret a form, you use reasonableness of data to validate the fields. For example, you know that a date field should have a certain format, you know that an address has certain characteristics. Validation performs the same checks on the forms data automatically. Without validation the forms processing system cannot guarantee accuracy. Validation of the data is implemented in many ways against

- edit pattern
- ranges
- validation fields
- possible acceptable characters
- check digits
- total counts or mathematical formulae
- look-up tables
- databases and other alternatives.

This is a whole industry in its own right and validations may be incorporated with the key validation repair process and /or run as a batch server operation.. Many of these validations are available as a part of the software forms processing systems. But to further assist, vendors are offering scripting languages and the ability to write one's own validations using Visual Basic.

Edit and Range Checks

This allows the user to set-up checks for patterns of characters. For example a date may consist dd/mm/yyyy. In this case the first field should be between 1 and 31, the second field between 1 and 12 and the third between 19xx and 20xx. Or it may be dd/mmm/yyyy, in which case the second field must be alphabetic. An amount should always have 2 decimal places, an age may be between 18 and 65 etc..

Validation Fields

Validation fields allow cross checks to other fields in the form. For instance if one field is completed then there may be a restriction on a check box being encoded or another field being completed.

Possible Acceptable Characters

Apart from those that are validated by edit checks, the system can check for accepatble characters. For example an amount cannot contain an alphabetic character. A person's name cannot contain a numeric field.

Check Digits

Check digits are numbers that are mathematically calculated from the contents of a field and added to the end of that field. By calculating the check digit from the recognized characters in the field and comparing that to the recognized check digit, the system can guarantee accurate conversion and if it is not correct place in the manual repair queue.

Counts or Formulae

Many fields are derived from other fields. An invoice for example has extensions to the line items which are calculated from the amount ordered times the unit price. There is then a total at the bottom of the invoice. Forms processing systems have the ability to make these calculations to verify the results of OCR.

Look-up Tables

Sometimes a field can be validated against a look-up table. One example of this is postal code look-up. A 5-digit zip code can validate the state and the town. A 9-digit zip code can validate a street. Some vendors are offering a DDE connection, others are offering ODBC connection and some are offering many different database connections.

Database Lookups

All forms based data must be interfaced to and connected to a database. One of the reasons people moved away from a batch data entry process was so that input could be validated against the on-line database and data could be extracted, reducing keystrokes. For example a customer number would retrieve the customer name and address, a part number would be validated and the associated description could be visually compared with the part description on the invoice. Alternatively, the partial description could retrieve a part number which then got keyed. All these functions require the use of high speed look-up tables. Some of these tables are massive and standard retrieval mechanisms are not always quick enough to keep up with a fast operator. This does not matter in the case of a batch process, but when an operator's rhythm is destroyed, productivity suffers badly. Specialized look-up mechanisms have been built to deal with tables that are hundreds of thousands of records long.

Image Based Key Repair

The output from the various recognition engines is placed in a file with its associated image(s). Post process look-up tables and validation tables may have increased the number of found errors through incorrect recognition or incorrect original data.

No forms recognition and processing will ever completely recognize and process the information on a form automatically. Tough words maybe, but accurate. Even if OCR is 99.99% accurate, it still means that 1 in 10,000 characters is known to be wrong.

If an average claim form contains 100 characters and a company is processing 30,000 forms a day, this means that 300 characters are known to be wrong. Then, substitutions, which are not known by the recognition engine, can add a further 1,000-5,000 or more characters. All this un-recognizable data and any substitutions must be key repaired. Furthermore, a blind key verification step must be carried out to ensure 100% accuracy.

When a recognition engine outputs its data, it does it on a character by character basis and calculates accuracy on a character basis. If this is 95% then 5 characters in 100 are known to be wrong. But, on the basis that the wrong characters are evenly distributed, then the numbers of fields that are incorrect will be 5 characters dispersed across 20 fields. This means that the operator must look at 20 different fields to key the repairs.

This is the manual part of the operation and at an average domestic loaded cost of $20/hr., can get expensive.

Heads-up key entry is when an operator looks at a split screen. The image of the form to key is located in one half and the key entry fields in the other half. Heads-up key entry has been invented to deal with mixed recognition and key entry, and a variety of vendors specialize in key entry. All have added or are adding heads-up key from image. Although one might think instinctively that keying from image must be faster than keying from paper, this is not always so. An average key operator can key at approximately 11,000-15,000 keystrokes/hour from paper and can pace him/herself to a rhythm assuming that the form layout is logical and the data easy to pick up and understand.

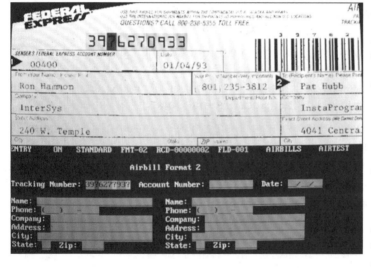

**Figure 13.1
OCR Repair
Heads-up
key entry**

Image

Data

*Courtesy of
Captiva
Corporation*

OCR repair can be categorized into three areas:

- high speed non-contextual repair,
- contextual repair
- repair with validation.

High speed non-contextual repair works on the basis of looking at the part of the image that is suspect and keying on top of it or next to it one character after another. No other data apart from the suspect character is displayed - this type of approach sacrifices all validation to ensure high speed. Success rests on the basis of how well the OCR engine can identify errors. If a substitution is not caught, then this system will not realize that keying is required.

Contextual repair involves identifying the character in the context of the word from a snippet of the image displayed on the screen.. If the operator sees the word STROET displayed, it is obvious that the wrong character is an E, but if the bitmap is displayed on its own then it may be difficult to ascertain that it is an E. Often contextual repair is paired with non-contextual repair. The operator rapidly enters all characters that he or she can identify as individual characters - any that cannot be immediately identified are tabbed over and placed in a "contextual" queue that is used to display the complete word.

Sometimes operators need to refer to a different part of the image to discern what the correct characters should be. In this case, the software usually provides the ability to move around and look at a different part of the image. Some vendors have implemented a hot-key approach to this that can be used for batches of the same type of form. In this case the operator needing to verify a field hits a dingle key depression which automatically shifts the image to the field that is needed for validation.

Repair with Validation

Repair with validation checks the characters in context. It allows the data entry field to be checked at the same time as the operator is entering the data .

Screen Design

Screens for data entry need to be designed to maximize operator productivity. Colors can be used to improve this if used carefully. Studies have shown that operators react well to the same colors as used in traffic lights.

- Green = OK, go ahead.
- Yellow = Warning but continue, possibly with a change ahead.
- Red = wrong, stop.

More than a very limited set of colors as above confuse and reduce productivity. Do not use a product with distracting exotic colors in the banner or instructions.

Screens must also be laid out in a logical order. The order of the data entry in relationship to the form display is also important in letting an operator's eyes track in a consistent manner.

Performance

Key operators who operate at an average of over 10,000 key depressions/hr per eight hour shift need to create a rhythm to maintain productivity. To maintain this the information has to be presented to them consistently, at constant speeds. Design of any key from image system must consider this. To be effective, images must be distributed across the network and OCR repair has to be able to keep up with the operators. Any validation must be in real-time without causing delays. It is important that any packages you select be capable of dealing with the volumes that you envisage and can access any database information without delay.

Keyboard Layout

A standard PC QWERTY keyboard is the most commonly used keyboard, with the 10 key pad numeric keys set up as shown below

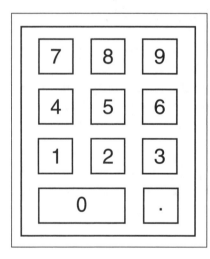

Figure 13.2
Example of standard PC keyboard numeric keys

but some operators key numeric data much faster using an 029 keyboard, so some software allows reconfiguration of a standard PC keyboard to an 029.

029 Keyboards

Back in the days of 80 column punched cards, IBM developed a specific numeric keyboard layout slightly different from the one used on today's PC's. This keyboard which became the standard for data entry and key-to-disk operations, differs from the PC's primarily in that it has different control keys; allows only for uppercase characters; and the numeric keyboard layout is different with the zero at the top center and 1-9 running from top left to bottom right (shown below).

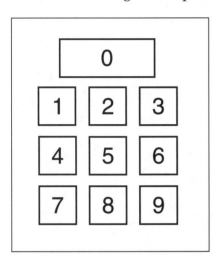

Figure 13.3
Example of "029" numeric keys

Many systems and high speed key operators were brought up with these keyboards, so they gain maximum productivity if the PC's keyboard can be re-configured as an 029. For this reason a number of vendors offer the ability to remap the PC keyboard to 029 keyboard functionality.

Verification Keying

Verification is the re-entering of data. Some fields MUST be 100% accurate regardless of cost. Whether the capture process is automatic, manual or a mixture of both, errors will creep in despite extensive validation checks. Vendors talk about 99.99 percent accuracy and this can be achieved in the first pass, but a user who is processing 10,000 or 50,000 forms a day may have to capture 2 million to 10 million characters of information 100% accurately.

99.99% accuracy still leaves 200 to 1,000 wrongly converted characters in the data. The only way to guarantee 100% accuracy is for a different operator to re-key this data over again.

High volume forms processing vendors provide the ability to re-key from image. The form is keyed blind without knowledge of what the first interpretation was, but in the background the keyed data is matched against the first pass of data. Any failures cause an interrupt, where the operator must review the data and rekey again.

CHAPTER

The Integration of Image Capture

Forms' processing was developed as a way to capture the data from forms more effectively. It used digital images of the paper forms as a basis to extract the data leveraging from character recognition and other recognition technologies to automate the process.

At the same time, companies were adopting digital imaging systems to convert their paper based forms to digital images, improving the storage and retrieval of them. This started as a simple microfilm replacement and the systems were known as file/folder systems as they sought to mimic a manual filing system. But then they expanded to the control and management of the paper electronically from the time it entered the company.

Initially these systems scanned a document, it was visually checked for image quality and the retrieval indexes entered. This is not an efficient way to operate since high speed scanning devices, which could convert a page in less than three seconds, were being constrained by the speed of the indexing process.

To solve this problem a variety of manufacturers developed batch scanning systems (see chapter 7). Initially these systems were based on inserting a number of different levels of coded batch control sheets in front of the document levels. E.g. folder, file and document. This increased batch preparation times but the improvement in scanning time easily off-set this. Still scanning is expensive typically costing between 6 and 10 cents per page (see fig. 14.1 for analysis of this).

Scanning Cost Calculation					Hours/day	8	
Equipment	. of Systems =	2	per annum	p. hour	Total		
Scanner	$ 30,000.00		$ 6,000.00	$ 30.00			
PC	$ 5,000.00		$ 1,000.00	$ 5.00	$ 35.00	$ 4.38	p.hr
Hi Res Monitor	$ 1,500.00		$ 300.00	$ 1.50			
QC Station	$ 2,000.00		$ 400.00	$ 2.00	$ 3.50	$ 0.44	p.hr
Electric	$ 0.15	KWH					
Watts/system	400						
Watts/day	6,400			$ 0.96		$ 0.12	p.hr
Space							
Sq ft/system	60						
Cost/sq ft/yr	$ 12.00		$ 720.00	$ 3.60	$ 3.60	$ 0.45	p.hr
					Total	$ 5.38	p.hr
Batch Scanning Calculations			Pages =	10,000			
Sheets/batch =	100						
Batch shts/bat	1						
Total batch she	100						
Tot. Sheets	10,100						
Seconds/page	2.75			7.72	Hrs		
Set-up time	15	mins		0.25	Hrs		
Jams	1	/hundred		101	pages		
Jam fix time	3	mins		5.05	Hrs		
Productivity	75%		Total Hrs	17.35	Hrs		
			Cost	$ 347			
QC and Rescan Calculations							
Display time =	2	sec/image		5.61	Hrs		
Percent rescan	7%			700	pages		
Secs/page	6			1.17	Hrs		
			Total Hrs	6.78			
			Cost	$ 136			
Batching Calculations							
Envelopes Pag	800			12.50	Hrs		
Counting	2000			5.05	Hrs		
				17.55			
			Cost	$ 315.90			
Equipment & Electric							
	10,000	Pages					
Scanning	$ 75.92				$ 0.008	per page	
QC & Repair	$ 2.97				$ 0.004	per page	
Electric & spac	$ 13.75				$ 0.001	per page	
	$ 92.64						
Labor							
Batching	1,000	sheets p.	$ 18.00	p. hr	$ 0.316	$/page	
Batch Scannin	10,000		$ 20.00	p. hr	$ 0.035	$/page	
Q. C. & Resca	700		$ 20.00	p. hr	$ 0.194	$/page	
Average Cost/Page					$ 0.09		

Figure 14.1
Cost of scanning can be substantial

But users wanted more indexes and they wanted to capture the data from the main forms. They wanted to search and find documents based on individual words and they did not want to re-scan. To satisfy this the document capture companies started to implement OCR and other recognition technologies at the document level. At the same time, the forms processing vendors whose images were often already being imported into back-end document storage solutions, started to capture the remaining documents even though there was no required data on them.

This has led to an overlap between the image capture vendors and the forms processing vendors. The two requirements are different but they are sufficiently similar in their processes and re-scanning too expensive that it is inevitable that the two converge.

Off-Shore

A major advantage of image based keying is that it removes the physical location requirement from the key entry process. Images can be transmitted to any part of the world in seconds and keyed from there using heads-up key from image software.

As a result imaging is liberating data entry from the physical locations close to the United States.

You might call this the second coming - the first occurred with the advent of the airplane which enabled users to locate their data entry operations in off-shore locations with lower labor costs than in the US. By the 1970's this had become a standard way of doing business.

The paper, or sometimes microfilm, was sent by courier to an international airport from which it was flown to a low cost offshore location. Once there, it was keyed onto magnetic tape, which was returned to the US usually with the original paper record.

Critical to success were:

- a time zone close to the US;
- an effective air service;
- an educated labor force that could understand English.

Sites sprouted around the Caribbean and Central America with some in Ireland and other European locations. But despite the low offshore labor costs, US locations in more remote and lower cost parts of the country were still competitive - hence Service Bureaus set up locations in remote parts of Pennsylvania, W. Virginia, Kentucky etc.

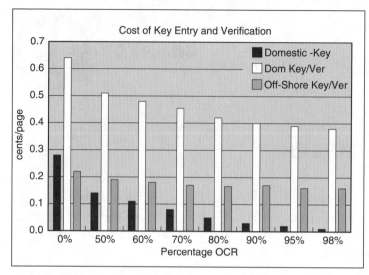

Figure 15.1
When combined
with OCR,
off-shore data
entry for repair
and verification
can be very cost
effective

As a result of the reduced communications costs and availability of images, we now have data entry Service Bureaus that have opened up in Bangladesh, China, India, Pakistan, the Philippines, Mauritius, Sri Lanka, Zimbabwe and other places.

Due to work knowledge and the continuing need for paper based processing though, many Service Bureaus are still located in Caribbean countries. These include Barbados, Belize, Dominica, Dominican Republic, Granada, Jamaica, and Trinidad where labor costs are as little as one tenth those in the US and the workforce is mostly English speaking.

Developing Countries Seek Data Entry

Data entry services are sought after by developing countries' governments. The work is clean and non-polluting. It does not require physical strength and the infrastructure costs are low. It utilizes high technology and trains local people in computer programming, communications, networks and other skills useful in today's and tomorrow's markets.

Document Imaging is a catalyst to expanding the off-shore market as it lifts any location barriers. With very high speed telecommunications lines running at up to one and a half megabits/second (T1 link), images of the paper can be delivered anywhere in the world within seconds. Data entry from these images can be carried out anywhere, turned around in hours and the physical paper never need leave the US.

In fact in some cases farther away can be better. Keying during the daytime shift is usually more expensive as the best operators work then. So if you can key during a day shift and turn the paper around overnight you gain two ways. For example by locating the key entry Service Bureau in China or India which is over 12 hours away from the US, overnight turnaround can be provided using the day time Prime Shift Operators who are normally priced higher.

As mentioned above, the paper never leaves the United States. It can even remain at the originating company's offices if they scan it there and with the growing use of document imaging systems for filing and storage, the costs of scanning are already absorbed.

However currently most forms are being scanned either at the gateway Service Bureau which maintains a high speed communications link, or at a centralized collection point. This has been necessary as the cost of the high speed leased lines used has meant that the pipe must be full — any reduction in data volume means that the transmission costs makes the transmission costs uneconomic.

However, this is changing with the Internet in particular being used as a low cost transport mechanism.

Once the paper is scanned the images can be transmitted to one or more low labor cost locations where it is keyed from the image and the data is then transmitted back to the originating location.

The advantages are numerous -

- the originating company can maintain control over the paper.
- the turn-around can be very fast regardless of location;.
- data entry quality and speed benefits from image based lead through and validation controls.
- jobs can be scheduled to meet the primary 1st shift even if turnaround is overnight.
- if necessary a job can be routed to more than one location to meet time constraints.
- OCR can be employed to provide automated control of batches and images as well as, in some cases, first level entry (see fig 15.1).

Location is Still Important

Even though location can be anywhere, it is important to understand where a Service Bureau may be doing the work. Care should be taken when choosing as it may have a major impact on five critical areas that affect the timeliness and quality of data entered.

These five issues are:

- Labor Cost.
- Security
- Telecommunications Cost and Image Sizes
- Language, Literacy, Education and Work Ethic of the Operators.
- Political Stability

Labor cost:

Direct labor figures are $1-$1.50 per hour in most of the popular off-shore locations (see figure 15.1). However, these figures cannot be compared directly with US figures as the government regulations regarding workers vary from country to country and have a direct effect on the total cost.

One example is that you must pay 13 months salary for 12 months work in China! Mexico also has work rules regarding vacation pay, layoffs, and other employer-supplied benefits. But generally costs are substantially lower than those in the US, where labor can vary from a low minimum wage to $15 or more.

Security:

One of the problems associated with offshore data entry is security of the data. You may not want your sensitive data delivered outside the US's jurisdiction. Tax and other personal information such as credit ratings need to be treated sensitively. But image based keying can help here too.

An image of a form can be easily cut electronically into two or more segments and keyed in by two separate people even in two different locations. The data can be re-assembled at the company that needs the information. You keep the key that puts the parts back together. Each person or location then only has access to a part of the document and without the other part(s) that information is useless. It's the same principle that cryptographers have used for years.

In fact, keying in this way, can easily be made more secure than it is using conventional methods with bonded labor.

Telecommunications Costs

Images are not small amounts of data. To run an offshore data entry service, using a direct leased line, a Service Bureau must use a 1.5MBPS T1 link which transmits data at about 200KBytes/second. For this privilege, you'll pay around $18,000 a month to Jamaica up to around $50,000 per month to China. A standard letter sized document scanned at 200 dpi typically takes up about 50K bytes of space. Therefore, a T1 link should be able to transmit 14,000 pages of data one way in an hour assuming it is an average page without image reduction (see below).

Leased T1 lines however have two problems:

- It is a fixed overhead.

You must keep the pipeline full to keep transmission costs low, so the effectively priced Service Bureau will be looking to saturate the lines. If a service bureau ran a theoretical 300,000 pages/day on a sustained basis 24 hours a day, 30 days/month using a T1 link to Jamaica it would cost just 0.2 cents per page to transmit.

- It must be provided from a centralized point

A Service Bureau collection position is normally the point of transmission to the place where key entry occurs. Paper must either be delivered to the collection location for scanning or images must be delivered over a fairly expensive high speed linkage.

ISDN links

ISDN (Integrated Digital Systems Network) may be an ideal mechanism to use to deliver images from small companies. Typically data entry needs vary according to the time of the day, month, quarter or year and effectively ISDN can expand in speeds to accommodate any volumes.

ISDN, consists of two lines each of which can carry data reliably at 64 kbps. This gives a combined speed of 128 kbps in the US, which means that 40-50K byte sized average sized images can be shipped at a rate of 15 per minute.

For low volume users who only need to send a couple of hundred pages a day, this represents a much cheaper option than traditional fixed speed leased lines. ISDN is now available at most locations in the US at a nominal rate. If a dial up connection is made, then the user only pays for the time used.

Therefore ISDN can reduce overall cost and it can be easily made available at branches or customer locations either to deliver images directly, or to feed a centralized collection system.

ATM

ATM is a new high speed protocol which may be very useful for image delivery. While the main specification is for very high speed transfer requiring dedicated lines, there is a new proposal to run at 25mbps over standard twisted pair lines. This would provide the average user the capability of transmitting approximately 50 images per second over a standard dial-up line.

The Internet

The Internet can be used in two ways for forms processing:

- In a corporate environment it can be used to deliver a small number of forms to any offshore location for the cost of a local phone call. Although delivery may not be immediate, the time delay is acceptable considering that a 24 hour turn-around is considered good. The data can then be delivered back to the sending location. As many companies have a T1 link to their POP (Point of Presence), the speed of the internet makes this a simple proposition.

- In a service bureau, the Internet can be used either as a collection mechanism from branches set up to collect the paper, scan it and return it to the customer. Or it can be used to transmit large volumes from a centralized service bureau who may have carried out functions such as security clipping or OCR/ICR to reduce the levels of manual entry.

Internet connections come in various guises from dial up to high speed lease line or cable modems. But even the lowest level dial up line running at a real 14.4 k bps, can transmit reasonable volumes of pages if the images are reduced in size and compressed (see later for a discussion on this).

Bitonal images are small compared to video or voice that is regularly being used over the Internet.

Reduction of Image Sizes for lower transmission costs

An average letter sized page image in gr/4 format takes about 50,000 bytes, but much of the information on a form does not need to be keyed. Data capture from image is different from image capture for management, filing and retrieval. Image sizes can be dramatically reduced by concentrating only on the data that is needed, then transmission costs fall substantially.

A secondary benefit for the service bureau is that smaller images represent faster throughput on a network so larger numbers of PC's can be carried on one server than would be possible in a standard client/server network.

Image Enhancement

The recognition pre-process including deskew, image clean-up and background removal will substantially improve compressed image sizes. Compression encodes runs of white or black lines or blocks. So if an image contains noise characters (speckles) these will cause bad compression. Likewise, skew can cause a run to be broken up into different lines making compression worse. Backgrounds, particularly shadings, will cause bad compression as there are many transitions from black to white and back again

There are three major ways to reduce image sizes:

1. Form Removal

This was discussed in the pre-recognition process chapter. This can reduce the image size to 15K or even less depending on how much information is on the form. On most forms though, the display terminals need access to a blank template that they can super impose on the image for context keying.

2. Image Segmentation

Much of the data on a form is often not required which can reduce image sizes substantially: examples are barcodes which have been interpreted, boilerplate text, photographs and other data that is not needed.

Separating the overall image into just the parts required for the entry of data, will in some cases dramatically cut image sizes.

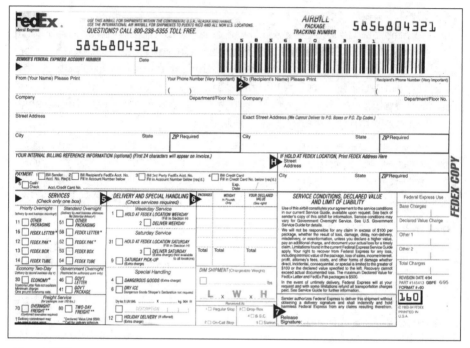

Figure 15.2
Required Data area is quite small.

Consider Federal Express' Waybill form, as an example. The data capture area - i.e. the area that's needed for keying - consisting of senders and recipients name, waybill number, shipping instructions/content, is less than 50% of the form. The overall size of the waybill is 47 sq in (approximately 23K bytes of image data when scanned at 200 dots-per-inch), but the capture data can be reduced by 16 sq in to 31 sq in (app 15K bytes). So around 48,000 pages/hour of data reduced, compressed Federal Express forms can be transmitted for the same cost as 14,000 compressed letter sized images.

A photograph on a form such as a passport application will cause large compressed images and it does not add to the captured data — in fact it may cause security worries, so use software to clip it out before compressing.

CHAPTER FIFTEEN

3. Transmit the Images at Display Resolution.

Probably an ideal solution for large color images, which provide the ability to avoid expensive rescans when the image is unreadable. A key entry operator can read an image for key entry purposes on a VGA monitor at around 100 dpi, assuming there is no very small text (i.e. that under 10 point). Any higher resolution is wasted. This reduces uncompressed image sizes by about three-quarters and compressed image sizes correspondingly.

Off-Shore Processing

The costs of data entry are usually measured in cents per key depression. The cost of scanning cannot be added unless there is perceived value by the customer in receiving archived images. In some cases this is easy, but in others where only the data is required, it is difficult. In these cases, the more data to be keyed per image size the more cost justifiable the off-shore operation becomes as the image transmission costs get absorbed.

Once the image is received at the off-shore location, it must be captured on the local server, identified, batched, routed and the fields that must be entered displayed. Any forms that need contextual keying must be matched with a display template. Verification — the need to re-key data that has already been captured

The language and literacy of the operators is critical. Many times an image display may not be entirely clear. This is particularly true if handwritten. An English competent operator can make an educated guess, one who only understands a foreign language will be confused.

Looking at the costs, the logic to move forms based data entry and particularly verification overseas seems overwhelming. Even taking a conservative view of loadings, image sizes and numbers of characters to key, the costs are attractive. The only other issues are related to the country.

- Is disruption due to political instability or trade friction likely?
- Is electrical power reliable and if not, are standby transformers available?

- Data entry has always been labor intensive. But automated forms processing supplemented with key entry from image offshore and delivery over the Internet reduces the costs substantially.

Professional Services: Surveys

Carlson Marketing Group Decreases Forms Processing Time and Data Entry Costs

Executive Summary

As a leading provider of global marketing services to businesses worldwide, Carlson Marketing Group (CMG) continuously researches new technology in order to deliver innovative services and maintain its competitive edge. In 1994, Nancy Nelson, operations manager at CMG, began to analyze the potential impact that an ICR/OCR automated data capture system would have on data entry costs, forms processing time, and CMG's operational flexibility. In late 1994, CMG decided to invest in the technology. In spring of 1995, the system went into production, achieving accuracy rates exceeding 99.99 percent.

Background

Some of the best known names in industry rely on CMG to manage their market loyalty and incentive programs. This involves collecting mountains of questionnaires and applications, entering the data, and keeping forms in accessible storage. The key to excellent service is quick processing. Equally important are the highest standards of data accuracy. CMG customers consider market information mission critical, making 99.99 percent data input accuracy an absolute requirement. The immediate objective for the system was to capture customer satisfaction data from six surveys, each containing 30 to 40 questions on double-sided forms, and archive the images for later retrieval in an image management system. The ICR/OCR/OMR front-end would be responsible for capturing the data and indexing the images.

The Solution

CMG's project moved into high gear when system integrators were brought on board to participate in needs analysis and system design. Working as a team, CMG and the integrators chose a forms processing software as a front end to an Optika's image management system. The forms processing software was chosen for its ability to meet three difficult challenges:

1. Data input accuracy had to be 99.99 percent or higher, indicating a need for sophisticated data validation and verification features to raise accuracy levels above that of raw recognition results.

2. Form re-design was not an option, which implied that CMG's solution would need robust image processing/form removal capabilities.

3. CMG did not want to add staff to convert from its existing technology. Simplicity and ease of use would be key considerations.

The Solution

Four ICR servers and three customized data verification stations are at the core of CMG's system. Once images are scanned to disk, the servers deskew images, automatically identify forms and proceed with form-specific image processing, recognition, and data validation. The servers run unattended 24 hours per day to ensure maximum throughput. Once the servers have done their work, data recognized with low confidence is assigned for daytime verification. All data is then subjected to a data validation pass, which can reroute character correction mistakes back to the verification step. Following verification and validation, data is uploaded to VSAM files on CMG's mainframe via a CIS program.

The same program that loads data to the mainframe simultaneously splits a portion of the data stream and uploads it, together with associated images, to an Optika image management application.

The Result

Today, the system is in production with flawless results. In a typical day, CMG's imaging system crunches through 20,000 images, automatically identifying forms and extracting the data. Data capture time has been significantly reduced, and, audits show data accuracy exceeds 99.99 percent.

"While it is too early to judge the total return on our investment, the system clearly represents a win for Carlson Marketing and for its customers," said Nelson. "We know that we are saving time and money on data capture and on questionnaire archiving and retrieval. What is more important is that the system has enabled us to better respond to customers' needs. Our data capture/imaging system allows us to accurately predict when jobs will be finished, quickly reprioritize processing as needed, and respond to clients' audit requests with the push of a button."

Adapted from a case study by Datacap, Inc.

Government: Census

Census Bureaus Move to Forms Processing Technologies

Governments around the world need to collect reliable statistics on their citizens in order to plan for future services and needs. Forms processing is a natural solution to automating the capture of this information and a number of countries have already adopted it.

Turkey

Turkey, as other member nations of the UN, was faced with the membership requirement of performing a National Population Survey. They needed a cost effective and time-saving solution for processing Turkey's 15 million national census forms.

On November 30, 1997, a national curfew was implemented from midnight Saturday until Sunday at 17:00. During this time approximately 600,000 teachers dispersed throughout the many provinces and villages of Turkey to distribute the census forms, complete the empty data fields, and return the completed forms to the regional manager, who delivered them to Turkey's State Institute of Statistics (SIS). Next was the task of capturing the data contained in the forms transforming it into usable statistical information.

After more than a year of searching, and a rigorous competition using a performance benchmark, SIS assigned the task to a systems integrator. Within two weeks the integrator constructed the entire facility to house the project processes, obtained all the necessary furniture and equipment, and recruited and trained over 600 personnel to participate in processing the data from the 15 million census forms.

The forms processing vendor modified the system to meet the government's need to:

- monitor the high volume of batches being transferred through the system,
- create a powerful statistics application which enables high-level ICR training to handle the Turkish alphabet consisting of characters previously untreated by any forms processing software, and
- provide on-line quality assurance services.

The system was comprised of eight segments. Each segment consisted of a Kodak 9500 scanner, an HP juke box (holding 128 MOs, or 338 gigabytes of information), one supervisor station and 20 completion stations.

After the scanning phase, the forms processing application generator, read, processed, stored and retrieved the information contained on the census forms. The system used multiple OCR/ICR/OMR recognition engines; sophisticated voting algorithms for weighting recognition engine results; and form recognition, removal, and image enhancement.

The system significantly reduced the time and cost, handling more than 600,000 forms per day for a total of nearly 20 million forms.

United States

The Census 2000 is the first U.S. population census to use automated forms processing software. Project planning has been underway since 1994 and by the time it is completed in 2003 it will have cost around $4 billion. The project will involve more than 8,000 operators handling 1.5 billion pieces of paper. Questionnaires will be collected and processed at four separate facilities across the United States.

For a project of this magnitude, the U.S. Census Bureau has appointed Lockheed Martin as the overall managing systems integrator.

Lockheed Martin will be working with a number of subcontractors and selecting software and hardware products from a variety of vendors.

China

Because agriculture is critical to the basic health of the Chinese economy – supporting 1.2 billion people, the Peoples Republic of China wanted to conduct an agricultural census of pigs, chickens, cows and other domesticated animal raised by the country's residents. A long lead-time for the collection of data would render the process almost useless. They needed an efficient way to collect and summarize data.

For the project, the government's data entry system would be primarily manual. Six million data collectors would fan out into the countryside to document the country's land and livestock. The completed census form would be sent into the data collection center in each province. The state statistical bureau in that province would enter the data and forward it to headquarters in Beijing.

For the 1997 Agricultural Census, government officials of the Peoples Republic of China wanted a solution for processing more than 240 million forms in 180 days with an error rate of .001, a dramatic shift in the process.

The Peoples Republic of China sought help from an international computer reseller specializing in mark sense readers (forms and hardware). Government officials decided to use the existing form, requiring the incorporation of hand print, or intelligent character recognition (ICR), technology into the solution. Several products were field tested before making a final decision.

The next step was the purchase of the software and hardware. The most critical function was the training, which was a two-and-a-half-week program, for the approximately 60 project leaders. When the leaders returned to their provinces, a box of equipment – PCs, scanners, and software – was waiting for them. They were then responsible for setting up the equipment, installing the software, networking the computers, and testing the systems.

The Peoples Republic of China collected and processed 240 million surveys in 180 days with an accuracy rate of 99.9 percent.

Adapted from case studies by Captiva Corporation, Microsystems Technology, Inc.and TiS,

Healthcare: Health Insurance Claims

Employers Health Insurance Leverages Technology

Background

Employers Health Insurance (EHI), a freestanding subsidiary of Humana Inc., is the 10th largest health insurer in the U.S., with annual premium revenues of $1.5 billion. EHI specializes in health coverage through managed care contracts, servicing more than one million members nationwide, with 18 sales offices and 3,100 employees. In addition, it provides dental, group life, long-term and short-term disability coverage; and administers state health insurance plans for California and Iowa. As a result, EHI must process more than 65,000 paper claim forms per week, an operation that has been automated since 1993 through the use of image-based, automated data entry software powered by intelligent character recognition (ICR) technology.

Growth Demands Efficiency

In 1993, Employers Health Insurance anticipated that their accelerated business growth would drive an increased demand for processing health claims more rapidly and accurately. Five years of growth witnessed a 30 percent climb in membership to 1.3 million members in 1998. At the same time, their medical claims submission rate jumped from 17,500 to 65,000 paper forms per week — growth that initially threatened to stretch EHI's processing capacity to the breaking point.

To manage this rapid growth, EHI added employees judiciously and leveraged data capture technology to reduce labor intensive jobs, helping to limit the number of new hires. Their major technology components include forms processing to automate data entry and help contain costs by reducing manual labor requirements.

Employers Health took four months to develop the forms processing software system internally before it processed its first live claim. The system includes one high-speed scanner, two ICR servers, and 13 reject repair workstations. The installation paid for itself in 18 months and, four years later, continues to produce a healthy return on the initial investment.

Because Employers Health Insurance's corporate image is very customer service oriented, quality of data capture has always been a priority. The company's goal is to settle and pay claims within seven to eight days of receipt. With three million online transactions, 30,000 electronic claims, 13,000 imaged claims, and roughly $5 million in checks to process daily, EHI has a vital need for high capacity, dependable technology.

Before automation, EHI was having difficulty obtaining and maintaining the proper staff to handle data entry growth. The work was being done manually or was outsourced through a service bureau. Driven by rigorous customer service demands, the mission of the scanning project team has always been to tightly control the parameters of cost, efficiency and timeliness by keeping their operations in house.

Using the resources of electronic imaging trade associations and by canvassing other industries, Employers Health decided that image-based ICR technology was reliable enough to automate their claims processing operation. Then they conducted a vendor survey to determine which products were the leaders in recognition technology. Based upon criteria that included usability, flexibility, scalability and experience in claims processing, EHI selected a processing software.

For application development, Employers Health wanted a software system that was readily accessible to their own developers. They found that most systems were proprietary, which meant they would have hire the vendor if they wanted to create new applications. EHI was drawn to the software because of its user-friendly application development. Lochmann observes, "Their product doesn't need a high degree of expertise to be effective."

Solutions for the Future

Given the complexity of health claim form processing, one of the most important capabilities that Employers Health saw in forms processing was the ability to handle "real world" health claims, the kind that don't rely on special inks or other devices to facilitate accurate character recognition.

Since the automated data entry software is modular, EHI has been able to add more forms processing capability to keep up with the ever-growing mountain of medical claim submissions. The software's scalability has provided the EHI scanning system with a migration path that anticipates and accommodates their growth requirements and the PC-based open architecture platform makes it easy to upgrade hardware as well as software.

"The healthcare industry must drive the cost out of the delivery system to achieve affordability and accessibility," states Roesler.

Adapted from case study by Captiva Corporation

Financial Services: Personal Income Tax

Berkheimer Associates No Longer "Taxed" by Forms Processing

Background

Berkheimer Associates is Pennsylvania's largest independent local tax administrator, processing tax records and payments for over 1,100 municipalities and school districts throughout the state. Pennsylvania residents pay a percentage of their individual earned income as a tax to municipalities and school districts. Municipalities and school districts, in turn, retain firms like Berkheimer as a designated administrator for the local earned income tax. In this role, Berkheimer receives every annual tax return addressed to its client municipalities or school districts. Berkheimer staff must open every envelope, sort each form by municipality, enter data, process data and funds, issue refund checks, complete data entry for its records, archive records, and make information readily available to its clients.

The Challenge – Paper Tax Returns

Despite recent IRS initiatives toward electronic tax filing, the bulk of Pennsylvania residents still file their taxes via paper forms. The paperwork load Berkheimer carries is enormous — and the growth in its business compounded the overload. Last year Berkheimer processed over half a million tax forms — with most forms containing three to four pages of attachments such as W2 forms, and many forms being joint returns filed by couples. This year, Berkheimer is printing over 1.2 million forms for distribution to taxpayers — a 15 percent increase from the previous year. Although tax return audit activity proceeds year-round, all tax refunds in the state must be issued by July 1 — creating a short time-frame between April 15 and July 1 for the initial data entry phase required for processing refunds.

Berkheimer added part-time staff during its peak season to handle the increased processing volume. Staff would manually re-enter municipal/school district tax information into Berkheimer's computer system. After manual data entry, the firm temporarily filed the forms until retrieval (if an audit was required) and then archived them. However Berkheimer knew it could decrease processing time and provide greater efficiencies throughout its process — reducing paper handling, filing and retrieval time, and also saving valuable storage space — with a forms imaging system. The firm's director of information services was familiar with information capture software and wanted to create a total imaging solution.

Solution

The company enlisted the aid of an outside systems integrator. Together, they assembled a system that far exceeded all expectations. Out of all forms scanned in the past year, 30 percent completely balanced and required absolutely no manual intervention; the software "did all the work." The rest of the scanned forms required some manual verification, correction, or amendment due to status or address change.

These efficiency improvements allowed Berkheimer to finish inputting total tax return data needed for its records a full three months earlier than in the previous year – despite an increased processing workload. The labor savings eliminated the need for additional staffing to process forms, despite the firm's rapid growth. Berkheimer found their forms processing software delivered increased efficiencies across the board that translated into money and time saved, better use of staff talents, and improved customer responsiveness.

Next Steps

Berkheimer increased the percent of forms it scanned into the system from 60 percent in the application's first year of use to 80 percent in the second year. And they plan to go further. Berkheimer also is working to modernize its forms as a way to fine-tune optical character recognition (OCR) accuracy and data verification.

Adapted from a case study by Microsystems Technology, Inc.

Government: Jury Summons

Massachusetts' Office of the Jury Commissioner Uses Software to Conquer Postal Blues

Background

Beginning in the mid-1980s, the Office of the Jury Commissioner (OJC) for the State of Massachusetts experienced a mail overload so severe that even the Commissioner himself was pitching in to open incoming jury summons responses.

"Everyone, and I mean everyone, in the office was involved," Gregory Fulchino, systems analyst for the OJC, remembers. "We had a staff of 20 during the day, and a nightshift of two to four temporary workers, and it still wasn't enough." During peak times, the night shift swelled to five or six, but even then, Fulchino and his co-workers regularly took mail home to open and sort during their off-hours. "It was a nightmare," he confesses.

The Challenge of a New Program

The mail increase was a direct result of the success of an innovative program called "One Day/One Trial," in which each citizen was eligible to serve as a member of the jury pool. Instituted in one county in 1979 to raise participation levels, the program was soon expanded to all 14 counties in the state.

"It's a great program, a model that has been studied by other countries, but our jury summons mailings jumped from 100,000 to more than a million once the whole state was included," Fulchino notes.

Each jury summons response needed to be identified, logged and entered into the system. This work was in addition to answering the juror inquiries that came in to the office. "The situation was getting unbearable. We were getting a lot of burnout cases on the staff," Fulchino recalls.

The Solution Begins

When Paul Carr took over as Commissioner, he launched an investigation to see if technology could ease the burden. By late 1985, the OJC had settled on a hardware-based scanning system. The system was huge by today's standards, filling a 10 foot by 10 foot room.

"The system cut a week off our turn-around time, and was a Godsend at the time," enthuses Fulchino. But soon a week wasn't enough.

Still More Progress

Locating replacement parts proved increasingly challenging and expensive. Maintenance costs ballooned to $10,000 a year. Another drawback was the system's inability to deal with any changes in forms design without extremely high program costs.

By 1992, Frank Davis was in office as Commissioner. Davis had a strong interest in the power of the PC. The OJC launched an extensive review that compared software-based and hardware-based systems. Eventually, a decision was reached that software-based systems would better meet the OJC's performance standards and offered more flexibility than available hardware systems.

A software system was chosen, and the task of writing code to perform the specific exception and handling issues processing tasks required by the OJC began. After installation, the number of juror responses processed each day doubled or better. From 1,500 forms, the OJC was now processing between 3,000 and 4,000 a day.

"We're still not testing this system's limits," admits Fulchino, explaining that there are many valid reasons for which a person can be disqualified, obtain a postponement, or have their court location transferred. These situations slow down processing time, because official acceptance of the prospective juror's reason for not participating is required.

Even so, Fulchino estimates that the remaining juror response cards (about 65 percent of the total) are processed with no operator intervention at all, something he admits to being "thrilled" about. He's also pleased about the system's flexibility. In the event that the OJC requires the redesign of the juror forms, the system can be easily adapted to accommodate the new forms with minimal effort.

Results

The system processes the 3,000 to 4,000 summons reply cards each day, supervised by only one verification operator. Because of the efficiency of the system, several staff members were redirected to other tasks that urgently needed attention and no nightshift workers are needed.

Fulchino explains, "We've cut costs, improved accuracy, and we're better able to arrange jury schedules. It's wonderful,". And the best part? "No one has to take mail home at night anymore," he jokingly concludes.

Adapted from a case study by Microsystems Technology, Inc.

Professional Services: Psychological Tests

McGraw-Hill/London House Uses Fax for Speed

Background

McGraw-Hill/London House is a leading developer of human resource assessments used by businesses and industries across the U.S. The firm, based in Rosemont, IL, has created more than 100 psychological test systems for screening and evaluating employees. These test systems are used by businesses that range in size from Fortune 10 firms to family businesses.

Challenge

To better serve its clients, McGraw-Hill/London House wanted to make it more convenient to score pre-employment tests and report results – a major service of the firm. The goals was to implement a system that quickly sent test answers to McGraw-Hill./London House for scoring and just as quickly sent comprehensive reports of test results back to employers. With a faster turnaround time, personnel staff at a client's office could administer a pre-employment test and then ask a potential employee to wait in the lobby while the test was scored and results returned. From the test results, the personnel staff could determine whether to continue with the screening process.

McGraw-Hill/London House's specific request was for a system that would accept faxed test answer sheets, score the tests and automatically fax back scoring reports to a company's personnel office in four minutes or less with no human intervention. Because the firm processed psychological tests, it was crucial to obtain a system that was highly accurate. In addition, the firms wanted a system that would identify faxing errors, such as document jams or pages missing, automatically capture the originator's fax number and immediately alert the sender so the problem could be remedied.

"It was important to us that we create a fax-based system," said Joseph Orban, Ph.D., senior director of product technology for McGraw-Hill/London House. "The fax machine is a general input/output device at companies large and small and it's easy to use."

The Solution

McGraw-Hill/London House sought the help of a systems integrator specializing in imaging and document management to develop a test-scoring system who provided

- data collection needs analysis,
- system design to meet specific needs,
- integration of the new systems with the current installed system,
- installation, and
- ongoing software support and maintenance, including a telephone help line and software upgrades.

The systems integrator recommended an automated forms processing solution which "met our goals of speed, accuracy and catching and handling faxing errors," said Orban.

The integrated document processing systems speeded, simplified and improved the data entry and document management processes. The system automatically scans or imports images from documents. It captures hand print, machine print, bar code and mark response and then interprets the captured information and translates it into ASCII data. The system automatically validates the data and transfers it directly into an existing application or document image optical storage retrieval system. Also, the system provides name and address contextual editing – another big time saver. Names and addresses are matched against a database of names and mailing addresses to check for errors in data entry, eliminating the necessity for validating data.

The systems integrator worked with McGraw-Hill to create a faxable form. "Many companies try using existing forms," said Orban. "I recommend throwing out any old forms and designing a new one from scratch." He attributes 60 percent of the success of the system to the new form created.

Adapted from a case study by National Computer Systems, Inc. (NCS)

Manufacturing: ESOP

Procter & Gamble Bets on Technology to Facilitate ESOP and Wins

Background

There's a good, a bad and an ugly side to consumer products' giant Procter & Gamble's (P&G's) decision to place a strong emphasis on employee stock ownership. The good side is that every one of P&G's 110,000 employees is given P&G stock as part of the company's One Share program. Employees not in current stock option plans were recently given a 100-stock option grant. Also good is that these purchase plans allow employees and retirees to "roll over" dividends into equivalent amounts of P&G stock purchased directly through the company and not through a brokerage, saving them brokerage costs and thus increasing the yields of their portfolios.

The bad is that most of these transactions are done through the mail, using paper-based forms. The system P&G implemented to process each application required intensive involvement from administrative staff and the workload was getting downright ugly.

Challenge

"Even though it's an electronic world, most of our transactions are done through the mail with a paper-based form," says Karyn March, systems administrator for P&G's shareholder services group. "Individuals call and request a new account application via an 800 number, and a mail-fulfillment service routinely sends out application forms that people fill out and return."

March notes the tediousness of a process where staff had to manually re-enter data written on each form into a PC, manually double-check each new account record, and manually roll up cash tallies to be posted each day to the company's mainframe-based system for stock transfer data. She adds that two staff spent nearly all of their 40-hour workweek processing the 500 new account-applications forms sent to the department weekly.

P&G knew that something had to be done to keep up with the growth the company anticipated without adding staff. "We needed a more productive way to work that still preserved or improved accuracy," March explains. But how?

Solution

Last June, P&G's March sought help, contacting a well-respected office automation and networking integration business for some suggestions. The company helped P&G to design a system, integrate hardware and interfaces to the software, and design forms that could readily be scanned.

Spectacular Results

The results were spectacular: P&G Shareholder Services group went from processing 20 forms an hour, to 120 forms an hour. By implementing scanners to eliminate manual re-keying of data, P&G slashed individual forms processing time from three minutes to 30 seconds per form, including the limited manual verification that is imperative for assuring accuracy on such cash transactions. The accuracy rate of data entered from the application jumped from 90 percent to 95 percent. And P&G saved more than 145 man hours each month, which freed one staff person for other tasks.

Next Steps

P&G plans to expand the application to include more automated forms, such as W9s for IRS tax status certification and forms for transferring certificates or shares to another person. Later this year, P&G also plans to add full document storage and electronic retrieval capabilities. These are steps that will help ready P&G for greater leverage of the Web in direct stock purchase process. While only 10 percent of new account applications originate on the Web today, this should expand as investors continually seek streamlined ways of managing their personal finances, March concludes.

Adapted from a case study by Microsystems Technology, Inc.

Service Bureau: Response Cards

Throughput Up, Costs Down!

PostLink Respons, a division of the Norwegian Post, is a service bureau providing packaging, insertions, retrieval of company address lists, and handling such as processing customers' response cards.

They installed forms processing software in March 1996, and during the first ten months, approximately 1.3 million documents were scanned and interpreted by the system. Installation was problem free.

"What we all found was that the system was quite easy to learn and that production got started very quickly," says Trine Aalborg, department manager at PostLink Respons. "We were really impressed by this."

Large jobs for PostLink processed 85,000 to 95,000 cards with both hand and machine-printed information.

"We got hold of the forms processing software through a synergy effect," says Aalborg. "STS — the State Computer Central — was using this software for its automatic data capture. When the Post bought this company, we discussed which systems and technology were worth retaining. This was definitely one of them."

Number one on PostLink's wish list is an address module, which can interpret the addresses automatically. "There is an enormous demand for this kind of service in the market," she adds.

"Of course, the bureau wants to extend the installation. Recently, they bought a new scanner, which they are extremely pleased with. This scanner produces very good images, which was crucial to a recent job that PostLink performed."

The forms had not been produced with automatic data capture in mind — the advertising agency had been given free reign. With the good quality of the scanner and the intelligence of the forms processing software, the work was done to everyone's complete satisfaction.

Summarizing her experience with forms processing, Aalborg says, "What we have gained is that we can offer much better and faster services to our customers. In their turn they will be able to reduce their costs even further. The profits have improved for all parties involved."

Adapted from case study by ReadSoft, Inc.

Government: Personal Income Tax

New Jersey Division of Revenue Cuts the Costs of Processing State Tax Returns

Background

In 1994, New Jersey State Treasurer James DiEleuterio, then assistant director of the New Jersey Division of Taxation, faced the burden of processing approximately 3.8 million personal income tax returns. His 300 data entry operators were keying returns on aging CMC/Unisys data entry terminals that could barely handle the existing volume of paperwork. He decided to automate the tax return processing system.

A major systems integrator won the job with a LAN-based forms automation solution using intelligent character recognition (ICR) software. The New Jersey Document Processing System (DPS) processes 95 percent of New Jersey's tax returns, which consist of New Jersey 1040 individual tax forms and HR (Homestead Rebate) 1040 forms.

The five percent of the returns that are rejected by the system as too hard to recognize are diverted for manual data entry.

Secrets of Success

According to Joe Roose, assistant director at the New Jersey Division of Revenue, the DPS project ultimately succeeded because New Jersey optimized the ICR technology. First came the issue of form design. The New Jersey forms were not "machine-friendly," so Roose had the returns redesigned to improve recognition performance. Special marks were printed on the forms to enable the machine to easily register the form for scanning. Data fields were repositioned on the tax returns so that a machine could easily locate them. Then his department changed the size of the envelopes so that the return could be folded only once, which resulted in faster document preparation and better scanning results. The developers of state tax filing software summarized the tax data in four columns that could be recognized accurately by machine. "We learned to say to the tax preparation software developers that we needed a 12-point Courier font, and to the printer that he should print in a specific PMS color, instead of just 'red' ink," said Roose.

The division also added graphic symbols to the returns which improved their ability to audit their work as the forms go through various processing stages. Roose reports, "We printed a different format bar code on each document within the tax return, which we use to check the completeness of the tax return and the order of documents."

Reengineering the Business Process

The technology required a reengineering of the business processes. After opening the envelopes, division staff sorts the returns into three categories:

1. Hand Printed
2. Machine Printed
3. Rebate Applications

Then they remove and process any attached checks. Returns are batched via a bar-coded batch control sheet in front of each tax return. The bar code is printed on the top, bottom, and sides of the batch control sheet so that staff does not have to pause to correctly position the form.

Each hand-printed submission consists of four sides with an average of four to five attachments, including W2s. Machine-printed submissions consist of 10 to 11 pages each. Automated processing starts with eight Kodak IL923 high-speed paper scanners each controlled by a Sun workstation. Scanning in portrait mode is quicker than in landscape mode. The W2 forms are problematic because they are all sorts of sizes and weights and have all types of printing on them. Bar-coded batch controls, which are recognized automatically, define the start and stop of each return. The bar codes on each page perform a check for completeness as well as a check on whether the paper was properly registered.

Feeds, Speeds, and Accurate Throughput

Once the forms are scanned, the images are routed in batches of 50 to the ICR engine, which is controlled by a UNIX-based, E4000 Sun server. Recognition of the data is handled through 14 Pentium PCs, interconnected through Ethernet cards, all of which run 24 hours a day during peak processing times. At an average of 350 to 400 characters per return, this represents DPS recognition throughput of about 1 million characters per day, per PC.

Field accuracy rates on the hand-printed forms are 90 to 92 percent, while the average is 97 to 98.7 percent for the machine-printed forms. On the basis that an average data field contains seven or eight characters, this equates to an impressive character-level accuracy of over 98 percent on hand print and over 99.6 percent on machine print. Human accuracy is 99.5 percent per character.

Accurate throughput is a product of more than careful scanning and excellent ICR. Contributing equally are rules and procedures that tightly define the characteristics of a given data field and that use ICR results to cross-check those fields against each other whenever possible. Their forms processing software has special programming tools that allow a user to easily set up complex data validation routines.

Thirty-three workstations verify data, then reconfigure it into the large record format that historically has been output from manual key entry stations. It is merged with exception items that have been keyed on ASCII-based screens and sent to New Jersey's mainframe for processing. After capture, the images are reformatted and passed onto a backend imaging system for storage and retrieval. Once the paper is scanned it can be boxed up, shrink wrapped and sent offsite for storage. Because all the data for a customer inquiry is available on the imaging system, the Revenue Department has never needed to access the paper storage area.

The Bottom Line

The forms processing software cuts the processing time per return by 75 percent and provides a better work environment. Last year the department processed 2.7 million returns and expects to complete more than 3.3 million returns this year. The number of workstations has dropped from more than 300 before the DPS system to 110. Savings are estimated at one million dollars annually.

Adapted from a case study by Captiva Corporation

CASE STUDY 10

Healthcare:
HCFA Forms

United Healthcare Corporation Processes 10,000 Forms per Day

Background

With 30,000 employees and projected 1998 revenues of $18 billion, United Healthcare Corporation is a national leader in healthcare management. United Healthcare has been serving consumers, managers and providers of healthcare since 1974. In order to reduce processing costs for the 10,000 HCFA claim forms they receive daily, United Healthcare uses automated forms processing software. By reducing the claim form turn-around time, automated forms processing software also improves United Healthcare's customer service.

The Challenge

Prior to implementing automated forms processing software, HCFA forms were received, sorted into batches, and microfilmed at a central inventory control location. The batched forms were then shipped to the appropriate processing department where claims processors entered the HCFA forms into the claims system. After entry and review, the forms were stored until destruction was authorized. If a paper copy were later required, inventory control would retrieve and print the document from microfilm. With this system, United Healthcare could process 55 forms per person, per hour. It was clear to United Healthcare that with the number of forms processed, any improvement in productivity would translate to a significant cost savings. Further, in the increasingly competitive healthcare industry, a more modern system could reduce turn-around time, improving customer service.

The Solution

United Healthcare selected a windows-based forms processing software solution that scans claims into the three-tier client/server Sybase application front end that United calls CPW. This front end also provides integrated imaging workplan management. The data is then transferred to their Unisys mainframe application COSMOS.

The Results

United Healthcare has calculated cost reductions of $0.163 per claim, primarily due to the reduction in staff time. Scanning a document takes less time than microfilming. Retrieving and printing forms directly from the Unisys mainframe is quicker than retrieving and printing the form from microfilm. And, of course the automated data entry saved time. United Healthcare reduced staffing levels at their inventory control center. Additional savings and improved customer service were realized through a reduction of approximately one day in turn-around time. United Healthcare intends to expand the program to include new claim types and to expand the use of automated claims processing to other health plans.

Adapted from a case study by Microsystems Technology, Inc.

Other Books We Publish

236 Killer Voice Processing Applications
ATM Users' Guide
Client Server Computer Telephony
Complete Traffic Engineering Handbook
Customer Service Over the Phone
Frames, Packets and Cells in Broadband Networking
The Guide to Frame Relay
The Guide to SONET
The Guide to T-1 Networking
Local & Long Distance Telephone Billing Practices
Newton's Telecom Dictionary
PC-Based Voice Processing
SCSA
Speech Recognition
Telephony for Computer Professionals
VideoConferencing: The Whole Picture

Telecom Books publishes books and magazines and organizes trade conferences on computer telephony, telecommunications, networking and voice processing. It also distributes the books of other publishers, making it the "central source" for all the above materials. Call or write for your FREE catalog.

Quantity Purchases

If you wish to purchase this book, or any others, in quantity, please contact:

Christine Kern, Manager
Telecom Books
12 West 21 Street
New York, NY 10010
212-691-8215
408-848-3854   1-800-LIBRARY
Facsimile orders: 408-848-5784

Telecom Books
6600 Silacci Way
Gilroy, CA 95020
www.telecombooks.com